• LOCOMOTIVES OF THE •
LONDON BRIGHTON & SOUTH COAST RAILWAY
1839–1903

EDITED BY JOHN CHRISTOPHER

AMBERLEY

First published 2014

Amberley Publishing
The Hill, Stroud
Gloucestershire, GL5 4EP

www.amberley-books.com

British Library Cataloguing in Publication Data.
A catalogue record for this book is available from the British Library.

ISBN 978 1 4456 3444 9
Ebook 978-1-4456-3451-7

Typeset in 10pt on 13pt Sabon.
Typesetting and Origination by Amberley Publishing.
Printed in the UK.

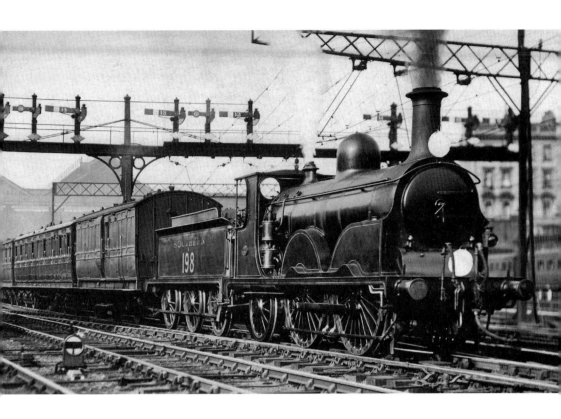

Contents

1903 Introduction   5

Chapter 1   Joint Committees and Amalgamations,   14

Chapter 2   LBSCR Locomotives, 1841–1849   18

Chapter 3   John Chester Craven, 1852–1869   39

Chapter 4   William Stroudley, 1879 – 1881   122

Chapter 5   Robert J. Billinton, 1890–1903   149

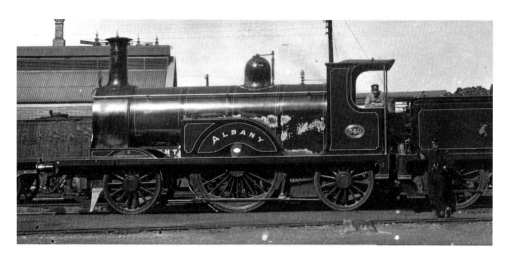

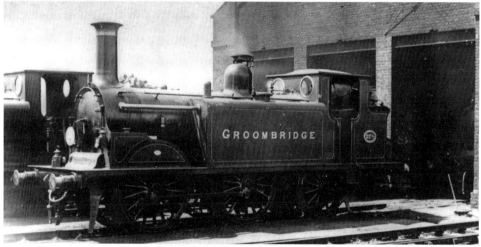

Two Stroudley designs: No. 349, *Albany*, the penultimate of twenty-six G class 2-2-2 engines built between 1874 and 1882. It was withdrawn in 1910. No. 278, *Groombridge*, is a D1 class 0-6-0 tank for suburban passenger services. built in 1879, withdrawn in 1926. (CMcC)

About this edition

Locomotives of the London Brighton & South Coast Railway is a revised edition of the book published in 1903 by the Locomotive Publishing Company. No author's name is given in the original which merely states that its contents were reprinted from *The Locomotive Magazine*. Therefore our thanks go to the unknown scribe whose work is being re-presented here for a new generation of railway buffs. The main part of the text is included, apart from some judicious editing to improve the flow for the modern reader, and all of the original line drawings have been retained.

This is the second in a new series of books from Amberley Publishing covering the locomotives of Britain's pre-grouping railways. **John Christopher**, *Editor*

1903 Introduction

So far back as the year 1825 an association was formed in London called the Surrey, Sussex, Hants, Wilts & Somerset Railroad Company, the object of which was to construct a line from London to Brighton, thence to Shoreham, Portsmouth and Southampton, and again to Warminster and the Bristol coalfields, through Salisbury, the capital of the proposed company being £1,400,000. Nothing more, however, was heard of this high-sounding undertaking, but others soon followed.

'Whereas the making a railway, commencing at or near Croydon in the county of Surrey, and terminating by a junction with the London & Greenwich Railway at or near Corbet's Lane, in the parish of St Mary, Rotherhithe, would be of great public advantage, by opening an additional, certain, and expeditious communication between the Metropolis and the town of Croydon, and the several intermediate and adjacent places, and also as a means of facilitating the ultimate establishment of railway communication between London and the Southern Districts of England, etc.' Such are the words which open the preamble of 'An Act for making a railway from Croydon to join the London & Greenwich Railway near London.' This preamble laid the seed from which germinated the whole network of railways now forming part of or connected with the system worked by the LBSCR.

The course of the line was indicated as starting on the western side of the high road from London to Croydon, near the ninth milestone from London, in the parish of Croydon, and passing from, through, or into the several parishes of Beddington, Streatham, Battersea, Lambeth, Camberwell, Deptford, Rotherhithe, Beckenham, Sydenham and Lewisham, and terminating by the aforesaid junction with the London & Greenwich Railway at or near Corbet's Lane. This pioneer work was duly opened on 5 June, 1839.

There is a general idea that the line from London to Brighton originally commenced near where Norwood Junction now is. This is quite correct, but running powers to and from London were always part of the plan, and it is certain that this point was never the Metropolitan terminus for trains coming from Brighton. One of the conditions under which the London & Croydon Railway (LCR) was sanctioned was that the company should buy up the site of the unprofitable Croydon Canal, dug in the clay and gravel soil through which the line passes from New Cross to Forest Hill. Between these stations there were some twenty-five locks; this difference in level accounts for the somewhat heavy bank on the railway of 1 in 100, known as the New Cross bank. The length of this incline is 2 miles 58 chains. West Croydon Station now actually stands on the original site of the basin of the old Croydon Canal, and portions of the old canal and towing paths are still visible at Anerley and in the neighbourhood. Although by the junction at Corbet's Lane, which was then the northern extremity of the LCR, passengers would have been able to continue the journey to a point near London Bridge, the directors thought there should be a separate and distinct Metropolitan terminus for their line, and accordingly an Act of Parliament was passed in the Session of 1836 enabling them to secure a site in the parish of St Olave, in the borough of Southwark, for the construction of a station, yard, warehouses, and other conveniences, the initial cost of which was estimated at £100,000.

About this time, the spirit of railway enterprise being very active in regard to the South of England, chiefly in consequence of the need of better travelling facilities between London and Brighton, a fierce Parliamentary campaign arose over various projected schemes. The Session of 1836 will long be memorable for the battle waged. Six schemes were promoted, the Brighton folks at first favouring one of the western lines, but ultimately adopting what was known as Rennie's or the 'direct' line. The lines proposed by the six engineers were as follows:

Stephenson's line – via Epsom, Leatherhead, Dorking, Horsham and Shoreham.
Rennie's line – via Streatham, Croydon, Merstham, Redhill, Horley, Balcombe, Cuckfield, Clayton and Patcham.
Cundy's line – via Mitcham, Epsom, Leatherhead, Dorking, Horsham, West Grinstead, Beeding and Shoreham.
Gibb's line – via Croydon, Ewell, Leatherhead, Dorking, West Grinstead, Steyning, Bramber and Shoreham.
Palmer's line – via Croydon, Godstone and Cuckfield.
Vignoles's line – via Croydon, Merstham, Horsham, West Grinstead, Henfield, Steyning, Bramber and Shoreham.

Except Stephenson's and Cundy's lines, all of them were intended to pass through Croydon. None of the engineers adopted the same terminus. Stephenson proposed to use the London and Southampton Railway, now the London and South-Western, from Wimbledon to its Nine Elms terminus near Vauxhall Bridge; Rennie's line ended at Kennington Common or the Oval; Cundy's at St George's Fields; Gibb's at London Bridge and Vauxhall; Vignoles's at the Elephant and Castle; Palmer's at London Bridge. The Brighton termini proposed had six different positions. The length of the proposed lines varied from a maximum of 54 miles in the case of Cundy's to a minimum of 47 miles in that of Rennie's. The estimated cost varied from £1,200,000 in Vignoles's project to £800,000 in those of Stephenson and Gibb; while the time to be occupied in constructing the line was placed by Stephenson at two and a half years, Rennie at three to four, and Cundy at two years. The amount of excavation and tunnelling required by the schemes was regarded with apprehension, almost amounting to dismay, although, of course, the construction of any of them would now be regarded as an easy achievement.

The company to whom powers of construction were finally conceded was compelled to buy up the old Merstham Tramway, which ran for a long distance parallel to the existing railway and close to the Brighton Road outside Croydon. The site of the old tramway is still plainly discernible to travellers in the Brighton trains, on the Up side near Purley Oaks. This tramway, whose full style and title was the Croydon, Merstham & Godstone Iron Railway, was one of the first of its kind. It ran from Croydon to Merstham chalk pits, and was joined at Croydon by the Surrey Iron Railway, a similar line starting from Wandsworth. They were worked only by horse-power. The rails, or tram-plates, were laid on small stone sleepers. Extensive remains of the first-named line may still be found by the Brighton Road side, parallel to the Merstham Tunnel. It was opened for traffic on 24 July, 1805. Part of the site was used for the new line.

At the time when the proposed railway to Brighton was occupying the attention of Parliamentary Committees, there were thirty-six coaches running between London and Brighton. These were collectively licensed to carry 3,453 passengers per week, and their actual weekly average in the year 1835 was 2,263, or a total of 117,676 per annum. The average fare paid by each passenger was 21s for inside and 12s for outside seats; the average time occupied on the journey was six hours. Two mail coaches ran daily, included in the thiry-six already named, carrying nine passengers on each journey.

However, at first the House of Lords rejected the Bill, in 1837, owing to the South-Eastern Railway (SER) having previously obtained an Act to construct a line to Dover *via* Reigate Junction or Redstone Hill, now Redhill. The Lords adduced as a reason that the SER already went 20 miles of the way to Brighton, and had been obliged to use this route; for when in 1836, an application was made for a line from London, the promoters of the company were entirely non-plussed by the action of the Speaker of the House of Commons, who remarked 'that no second outlet for a railway would be allowed to the South.' In 1837, Captain R. Alderson, R.E., being deputed by a Parliamentary Committee to decide which was the best of the four schemes then contending for an Act, reported on 27 June in favour of Rennie's, as being the shortest route and giving the most convenient terminus at Brighton. The Bill for this received the Royal Assent on 15 July. The termination of this Parliamentary struggle was that the Brighton Company was forced to make a line from Norwood, near the present junction, to Brighton through Redhill, and to sell, transfer, and relinquish one half of the line 'as shall be at or to the northward of the point at which the junction of the said SER with the said London and Brighton Railway shall be so authorized ... so that they may henceforth form a part of the SER and not of the said London and Brighton Railway.' The consideration for this transfer was, ultimately, the repayment by the SER to the Brighton Company of all moneys, costs and expenses connected with the works, together with the interest of £5 per cent per annum, making a total of £340,000. This jointly-worked line is 12 miles 5 chains in length.

So much for the circumstances leading to the South-Eastern Company's presence at Redhill. Owing to the constant delays that occurred at Redhill Junction from the traffic of the two lines uniting at that point, the Brighton Company have now constructed a new line between South Croydon and Earlswood. This new route relieves them from the hindrance to traffic caused by running over the same rails as the SER trains, but involved the reconstruction and rearrangement of East Croydon Station and the widening of the line from that station to South Croydon.

The new line begins at South Croydon and runs, as a widening, parallel to the existing main line on the west side to near Coulsdon. Its length is 3 miles 5 furlongs and 110 yards. It proceeds from Stoat's Nest southwards on the western side of the existing lines, and in order to avoid interference with the Coulsdon Station of the SER, diverges near it and is carried in a covered way and across two heavy bridges over the Brighton main public road. It then crosses the jointly-used SER and Brighton lines by a long bridge to the eastern side, passes through a deep chalk cutting, and a long tunnel through Lord Hylton's property, crosses Holmesdale Valley, and goes under the Tunbridge line (SER), by tunnel and covered way to a junction with the Brighton main line between Redhill and Earlswood stations. The length of this section is much longer – namely, 6 miles 5 furlongs, making a total of 10 miles 2 furlongs 110 yards.

Let us now return to the formation of the old line. The line from London Bridge to Corbet's Lane Junction was the property of the London & Greenwich Railway (LGR), and from Corbet's Lane Junction to a little beyond Norwood Junction (formerly called the Jolly Sailor Station) was the property of the LCR. To both of these companies the Brighton Company had to pay a toll of 3*d* for each passenger carried over their lines. In 1842 the line was widened from Corbet's Lane Junction to London, in consequence of the great increase of traffic due to the line being used by each of these companies. On the jointly-used line the trains belonging to the Brighton Company stop at the stations on the northern half, and those of the SER at the stations on the southern half of the line, but Purley (formerly Caterham Junction), although on the Brighton Railway Company's portion, is used by the trains of both companies. It was originally an exchange station for the Caterham Valley Branch belonging to an independent company, but an arrangement was contracted between the two companies for the joint use of the station.

The original gauge of the Brighton line was 4 feet 9 inches, i.e., the usual gauge laid ½-inch 'wide.'

In the year 1845 the Croydon Company laid, as an addition, a single line of atmospheric railway, a short description of which may be interesting.

A large tube was laid down in the centre of a line of rails. This tube had an opening at the top, which was closed by a valve formed of a leather strap covered with short plates of iron above and below. The strap being broader than the plates, it pressed against the top of the pipe by a succession of long rods, screwed down with hook bolts, and thus forming a hinge. On the side on which the valve opened was a groove, which was filled with a composition of wax and tallow. When the valve was raised there was space enough for a bent plate of iron to pass into the tube. To the leading carriage of every train a piston was attached by a bent plate; on the piston rod were four wheels, two before and two behind the bent plate, so that when the piston was in the pipe these wheels raised the valve and prevented it from touching the bent plate. In this manner a communication was made between the piston in the pipe and the leading carriage. Engines, working large air pumps, were stationed at intervals along the line, by means of which the pipe was exhausted of its air. When this was done, the pressure of the atmosphere on the back of the piston was the power to draw the train. The size of the tubes on the London & Croydon Railway was fifteen inches, internal diameter, and at a trial trip with a train of twelve carriages the remarkable rate of 75 miles an hour was said to have been attained.

After this arrangement had been in use about twelve months it was discarded, having proved totally unsuccessful. The old stationary engine-house still remains, however, at Forest Hill. Towards the end of 1845 a proposal was made to amalgamate the L&SW, Croydon and Brighton railways under the title of the Great South of England Railway, but the scheme fell through.

To return to the commencement of the Brighton line, the Bill for its formation received Royal Assent on 15 July, 1837, and on 12 July, 1838, ground was first broken at the deepest part of the great cutting on the north of Merstham Tunnel. The formation of the line and the construction of the works connected with it were carried on with the utmost vigour, and by a singular coincidence, the railway was opened for passenger traffic to Hayward's Heath on 12 July, just three years afterwards. The first permanent rail of the London and Brighton line was laid at Hassock's Gate on 4 February, 1839. Meanwhile, over a mile of permanent rail had been laid on the Shoreham Branch at Portslade, and other portions of the work completed. On 27 May, 1839, the foundation stone of the New England viaduct was laid with Masonic honours.

The first portion of the company's system opened for traffic was between Brighton and Shoreham, the event taking place on Monday, 11 May, 1840. The first train started at 3.00 p.m. The trip occupied 11½ minutes, the return journey taking a trifle longer.

Within a few days of the opening, a fatal accident occurred. A young man, incautiously sitting on the tailboard of a luggage waggon, which had been temporarily used to accommodate the extra traffic at Shoreham, was precipitated by a sudden jerk beneath the train at Southwick, and killed on the spot. The third-class carriages used were of the poorest description, little better than cattle trucks. They were wholly uncovered, and some had not even the accommodation of seats, the divisions of the sections in each carriage being simply an iron rail. Travelling under such circumstances was the reverse of pleasurable, the dust and smoke from the engine being annoying in the extreme. Worse than these was the almost constant descent of fine ashes, and umbrellas were in frequent requisition by passengers to prevent them getting into the eyes and to reduce the risk of getting burned.

Their condition when these unpleasant concomitants were associated with a high wind or a driving rain may be better imagined than described. By and by, 'covered carriages' were introduced, but without windows. These latter luxuries with, in some cases, cushions, etc., are modern improvements which were altogether unknown to third-class passengers in the earlier days of railway travelling.

As already stated, the London and Brighton line was opened on the 12 July, 1841, as far as Hayward's Heath – the passengers being conveyed the intermediate distance to and from Brighton by coach. The journey was thus accomplished in about four hours, four trains running from London and four from Hayward's Heath daily, and two each way on Sundays.

The whole line being completed between London and Brighton, the arrival of the first train with the directors and their friends, at noon on Tuesday, 21 September, was hailed by a triumphant chorus of all the musical force available in the town, assembled at the Brighton terminus. The Union Jack was hoisted at the Parish Church, the bells from the old tower sending forth a joyous peal. All was bustle and excitement, and numbers flocked to the terminus to witness the departure of the first train to London. At noon the great event of the day – the arrival of the first Down train with the directors and their friends – was to take place. At this time the terminus and all the points of vantage in the neighbourhood presented a most animated spectacle. Tickets of admission having been liberally distributed, many hundreds of people quickly assembled within the area of the terminus to await the coming of the train. This having duly arrived, the National Anthem was played and sung whilst the directors and their friends alighted.

The first 'Departure and Arrival Time-table' between London and Brighton in 1841 presents a marked contrast with that of the present time, both as to the number of trains and the names of the various stations. The trains were as follows. The term 'mixed' was applied to first and second class trains, which in general stopped at all stations.

LONDON TO BRIGHTON. – Mixed train, 9.45 a.m.; 1st class train, 10.45 a.m.; mixed train, 1.45 and 2.45 p.m.; 1st class train, 3.45 p.m.; express train, 4.45 p.m.; and a mixed train (London and Croydon), 7.00 p.m. Sundays: Mixed trains at 8.00 and 10.45 a.m., and 7.00 p.m. London to Croydon, 10 p.m.

BRIGHTON TO LONDON. – Mixed train, 6.45 a.m.; express train, 8.30 a.m.; 1st class train, 10.45 a.m.; mixed train, 11.45 a.m.; 1st class train, 2.15 p.m.; mixed train, 4.00 p.m.; and a mixed train, (Croydon to London), 2.45 p.m. Sundays: Mixed trains, 7.45 a.m., 4.00 and 7.00 p.m.

The stations were: Brighton, Hassock's Gate, Hayward's Heath, Balcombe, Three Bridges, Horley, Redhill, Merstham, Stoat's Nest, Godstone Road, Croydon, New Cross and London Bridge. Of these, Godstone Road was where Purley Station now is.

At the present time there are 23 trains daily from Brighton to London, of which 10 are express; from London to Brighton there are 27 trains, of which 12 are express. When the line was opened the fastest booked time between London and Brighton was 1 hour 45 minutes, and at the present time the journey is done in 1 hour on Sundays by the 'Brighton Limited' Express, Pullman car train. After these brief details concerning the early days of the railway, it is necessary to give some particulars bringing its history down to the present time.

The Coast lines were next simultaneously developed. On 24 November, 1845, Worthing received its first railway passengers (the west line, as far as Shoreham, having been finished as early as May, 1840), and Chichester on 8 June, 1846. On the same day the first train from Brighton reached Lewes, and before the end of the month the East Coast line was opened to Hastings. The session of 1846 witnessed the most important event that had yet taken place in the history of the line, *viz.*, the amalgamation with the Croydon Company and the adoption of the present comprehensive, if rather long-winded, title. The Act was passed on 27 July, 1846, the authorized capital of the combination being £4,730,000, with borrowing powers to the extent of one-third of that sum. In another twelve months (June, 1847) the West Coast line was extended to Portsmouth, and thus the three great arteries of the LBSCR system were completed. Including a branch from Croydon to Epsom, the extent of the line was some 140 miles, the total cost of the works up to that time being about £2,500,000.

Between October and December, 1847, the Keymer to Lewes and Lewes to Newhaven lines were completed and opened. By these a most important object was effected; the Port of Newhaven was brought to within 56½ miles of London, enabling the company to open a line of communication, by rail and boat, between London and Paris. The journey from one capital to the other was then made in from eleven to twelve hours. At the present day it is accomplished in from eight to nine hours.

A few months later (14 February, 1848), the line from Three Bridges to Horsham was opened. The branches to Hailsham and Eastbourne were opened in the following year (14 May, 1849), making a grand total of 187 miles available for traffic. Early in 1852 a dispute arose between the LSWR and the Brighton Company regarding the working of the Portsmouth joint line. The Brighton Company considered that the South-Western had taken an undue advantage of them, and, in retaliation, put on a steamboat service between Newhaven and Jersey.

The LSWR replied to this competition by promoting a line from Wimbledon to Croydon. Later on in the year the two companies patched up their differences, and, indeed, a proposal was again mooted to amalgamate their systems. Nothing, however, came of it, but in 1853 they arranged to construct a joint line from Wimbledon to Croydon. This line was opened on 22 October, 1855. For about half a mile on each side of Mitcham Junction it occupies the site of the old Surrey Iron Railway, before alluded to, which was opened for traffic on 26 July, 1803, and finally closed on 31 August, 1846. Soon after that the materials of it were sold. A branch of it went down Tramway Path to Hackbridge Mills. In 1856 the LSWR left off using the Croydon line, the Brighton taking it on lease; while from 1 January, 1866, they took it over altogether.

A company was formed in 1853 to construct a railway from Havant to the LSWR at Godalming. By 1858 it was ready for opening, but the company had made no arrangements for working it, nor had they provided any rolling stock. At the Godalming end they had a junction with the South-Western, but that company at first would not work the line, so the Portsmouth Company sought powers to construct a junction with the SER at Shalford. This the latter company opposed, alleging that its arrangements with the LBSCR precluded it from entering into any agreement with a competitor of that company. Parliamentary sanction was, however, obtained for the Shalford Junction. The line then only extended as far as Havant (8½ miles from Portsmouth) so, in 1858, powers were obtained for an independent line from Havant to Hilsea, parallel with the LBSCR line, with running powers over the joint line to Portsmouth from Hilsea (Portcreek).

The House of Lords, however, inserted a clause that the Portsmouth Company was not to use the Portsmouth Joint Station without the consent of the Brighton and South-Western Companies. Powers were, therefore, obtained for an independent terminus at Landport. In the summer of 1858 the Portsmouth Company agreed to lease their line to the LSWR for ever. This arrangement brought the Brighton and SER into conflict. Both now made active preparations for the fray. The following briefly relates the circumstances: The LSWR advertised that on 1 January, 1859, their new line to Portsmouth, *via* Petersfield, would be opened, but anticipating some friction between themselves and the Brighton Company owing to the latter disagreeing as to the entrance of the LSWR trains into the joint station at Portsmouth, they began running on 28 December, 1858, to enable them to get their service into working order by the New Year. They announced the arrival of their first train at Havant, the junction with the Brighton line, at 10 a.m., but with the idea of catching their rivals napping, they actually reached the junction at 7 a.m. The Brighton officials, however, were up too early, for before the LSWR train arrived they had taken up the rails at the junction, and placed an engine on the crossing, with the wheels chained to the metals and securely padlocked. After a scrimmage these difficulties were overcome, only to find others further on, and finally the LSWR train had to retire, and was sent back to London, whilst the LBSCR train returned to London, the locomotive being profusely decorated with rosettes. The battle was then transferred to the Law Courts, and finally on 24 January, 1859, the first through SWR train arrived at Portsmouth *via* the new line.

During the next eleven years attention was turned to the construction of lines connected with the Crystal Palace and the West End of London. The extensions from the Palace reached Battersea in March, 1858, and in October, 1860, they were completed across the Thames to the Grosvenor Canal Basin, Pimlico, at which point the company's West End terminus is situated.

But while the main line was thus improved and extended, the country districts were not neglected. The East Grinstead line was completed and opened in July, 1855; that to Uckfield in October, 1858; whilst in October, 1859, the Horsham line was extended to Petworth. In July, 1861, the Mid-Sussex line was opened to Partridge Green, and in the following September to Horsham.

After 1861 many additions were made to the lines at the London end. The most important of these was the company's participation in the West London Extension line from Clapham Junction to Kensington, which was opened in March, 1863. This branch is the joint property of the Brighton and other companies, and is the connecting link with several lines north of the Thames, so that a passenger may proceed from any station on the LBSCR to any station on the North London, Metropolitan, London and North-Western, Great Western, or District Railways.

In August, 1863, the line from Pulborough to Ford was opened, and during the same month the branch from Ford to Littlehampton. By the opening of this line the company secured, in connection with Littlehampton Harbour, an improved route for the traffic of the Channel Islands, and opened up a considerable traffic between that harbour and Honfleur and the coast of Normandy, to which it is the nearest port. This service was transferred to Newhaven some few years ago. Bognor received its first visitors by railway on 1 June, 1864, and on the same day was opened the extension from Newhaven to Seaford.

From 1864 to 1879 great strides were made in the development of the company's lines. In May, 1865, the branch from Sutton to Epsom Downs was opened, a length of a little more than 4 miles, and in the same month the West Croydon and Selhurst 'Spur', a connecting link between West Croydon and Victoria, was used for the first time.

In August, 1866, a portion of the South London line was opened for traffic, viz., from London Bridge to Brixton, the remaining portion being completed by May, 1867, to Battersea Park, thus forming a new short route from London Bridge to Victoria.

In 1867–1868 the Mid-Sussex line, *via* Dorking to Portsmouth Town, was opened, obtaining for the company a share of the Portsmouth and Isle of Wight traffic. The extension to the harbour was opened in 1876, and in 1880 the LBSCR, jointly with the SWR, decided to take over the steamers running from Portsmouth to Ryde and greatly to improve the service. They also rebuilt Ryde Pier, running a double line of railway upon it, so that passengers could step direct from the train to the steamer, as they were able to do at Portsmouth Harbour. The new pier opened for traffic on 12 July, 1880.

In December, 1869, the first portion of the East London line was opened, and in April, 1876, the remaining piece from Wapping to Shoreditch, whence the Brighton Company obtained running powers to Liverpool Street.

By an agreement dated 29 March, 1877, the South-Eastern Company is entitled to a share of the profits from the Eastbourne traffic, and some years ago used to run two trains each way (1st and 2nd class only) between London and Eastbourne, *via* Tunbridge Wells, Heathfield, Hailsham and Polegate; but after the service had been in operation about a year it was withdrawn. Instead of it the LBSCR now pay the SER a percentage of the receipts, usually amounting to about £29,000 a year, not to interfere with the Eastbourne traffic.

In April, 1880, a portion of the new Tunbridge Wells and Eastbourne line was opened, *viz.*, from Hailsham to Heathfield, and the remainder a few months later, from Heathfield to Groombridge. This section opened up a new route from Kent and East Sussex to the South Coast. In the following year, 1881, the Chichester and Midhurst section was opened. Other important sections that followed were the branches from Culver Junction to East Grinstead, Horsted Keynes to Hayward's Heath, South Croydon to East Grinstead, Oxted to Groombridge, the Dyke Railway, and the widening of the main line to Earlswood, which we have previously described. At the present time the company are widening the line from Victoria to Croydon, and are about to enlarge Victoria Station on a scale which will make it one of the finest termini in London. A few years back a station was built at Gatwick for the accommodation of race traffic. From Horley to Gatwick there are four sets of metals, which are being continued as far as Three Bridges. Next session, the Company will apply for Parliamentary powers to double the line through to Brighton, they also will seek power to electrify the system, should it be considered expedient to do so.

With these facts we conclude our notice of the history of the line. The table on two following pages shows the company's system of lines up to the present time, with their lengths, dates of authorization, and of opening for traffic.

The following sections of the LBSCR system are single lines, and are worked under electric train staff regulations: Culver Junction to Horsted Keynes, Lewes Road and Kemp Town, Stammerham Junction and Peasmarsh Junction, Barnham Junction and Bognor, Chichester and Hardham Junction, Three Bridges and Ashurst Junction, Polegate and Redgate Mill Junction; Newhaven Harbour and Seaford; Dyke Junction and the Dyke, West Croydon and Mitcham Junction, and Mitcham and Merton Park.

With this brief outline of the development of the railway we must now conclude, and proceed to the history of the locomotives of the LBSCR in the following chapters.

Chapter 1

Joint Committees and Amalgamations, Locomotive Stock 1839–1841

Previous to 1 March, 1844, the Brighton Company possessed its own locomotives, but from that date the rolling stock came under the joint control of the London & Croydon, South Eastern and London & Brighton Railway Companies, and was used in common by them all. Until the formation of this joint locomotive committee the Brighton Company's engines worked the trains of the South-Eastern Company to and from Redhill and London.

The following is a list of some of the original Brighton engines previous to the amalgamation:

Engine No.	Name	Date	Maker	Maker's No.	Cylinders	Driving wheels
–	Merstham	1839	Sharp, Roberts	44	14 x 18	5 feet 6 inches
–	Coulsdon	1839	Sharp, Roberts	54	14 x 18	5 feet 6 inches
6	Venus	1839	Sharp Roberts	45	12 x 18	5 feet 6 inches
10	Jupiter	1839	Sharp Brothers	70	12 x 18	5 feet 6 inches
11	Mars	1840	Sharp Brothers	101	14 x 18	5 feet 6 inches
12	Saturn	1841	Sharp Brothers	113	14 x 18	5 feet 6 inches
13	Mercury	1841	Sharp Brothers	134	14 x 18	5 feet 6 inches
14	Orion	1841	Sharp Brothers	136	14 x 18	5 feet 6 inches
15	Sirius	1841	Sharp Brothers	137	14 x 18	5 feet 6 inches
16		1841	Sharp Brothers	142	14 x 18	5 feet 6 inches
17		1841	Sharp Brothers	148	14 x 18	5 feet 6 inches
18		1841	Sharp Brothers	150	14 x 18	5 feet 6 inches
19		1841	Sharp Brothers	153	14 x 18	5 feet 6 inches
20		1841	Sharp brothers	154	14 x 18	5 feet 6 inches
21		1841	Sharp Brothers	158	14 x 18	5 feet 6 inches
22		1841	Sharp Brothers	159	14 x 18	5 feet 6 inches
–	Brighton	1840	G. & J. Rennie		14 x 18	5 feet 6 inches
–	Shoreham	1840	G. & J. Rennie		14 x 18	5 feet 6 inches

The first two engines appear to have been built by Sharp, Roberts & Co. in 1839, and were named respectively *Merstham* and *Coulsdon*. They had single driving wheels 5 feet 6 inch diameter, leading and trailing wheels 3 feet 6 inch diameter, cylinders 14 inch diameter by 18 inch stroke, and weighed empty 12 tons 10 cwt. The boiler and firebox dimensions were the same as those of the engines supplied to the Croydon Company by the same firm, the heating surface being of tubes 467.28 square feet, and of firebox 46.62 square feet. These engines were at first the property of the contractors who made the line to Brighton, but when this was completed they were purchased by the company.

In the same year, 1839, Sharp, Roberts & Co. supplied two engines similar to the previous ones, only that they had smaller cylinders; the diameter was 12 inches and stroke 18 inches. They were numbered and named 6 *Venus* and 10 *Jupiter*. *Venus* afterwards became No. 51 on the SER, and later No. 10 on the LGR. In 1840 they supplied another engine of the 'Coulsdon' class, No. 11 *Mars*, afterwards No. 55 on the SER.

In 1840 G. & J. Rennie built two single-wheel tender-engines, named *Brighton* and *Shoreham*, which worked the Brighton and Shoreham branch, the dimensions of the cylinders and wheels being the same as in the three engines previously mentioned, the boiler being 3 feet 6⅝ inches by 8 feet 3 inches, having 99 tubes of 2⅛ inch diameter, and 4 of 1⅝ inches, with a heating surface of 488.09 square feet, and the firebox 3 feet 1 inch by 3 feet 7½ inches, with a heating surface of 69.74 square feet, and their weight was 14 tons empty and 15 tons 12 cwt in working order. They were afterwards sold to contractors.

In 1841, Sharp supplied eleven more 14 inch by 18 inch singles, Nos 12–22. Nos 12–15 were named respectively *Saturn*, *Mercury*, *Orion* and *Sirius*, these four afterwards being Nos 56–59 on the SER. This closes the list of engines we can trace as existing before the amalgamation.

In July, 1846, the London & Croydon and London & Brighton Railways amalgamated, and became the London, Brighton & South Coast Railway (LBSCR).

The joint locomotive committee was dissolved on 31 January, 1846. The Brighton and the South-Eastern afterwards shared the locomotives between them, each retaining the majority of those which were formerly its respective property, and taking over a few which had belonged to the London & Croydon Company. After the committee was dissolved the Brighton Company's engines had a letter B painted on the buffer beams, so that when a train was approaching a station the staff there were enabled to tell to which company it belonged. At the present time an old tender of Sharp's stands in a siding at Brighton having B62 painted on the back, this not having been painted over since that time. The SER were allowed to use the New Cross shops for a year after the dissolution while their works at Ashford were being constructed.

The number of locomotives possessed at this time was fifty-four, but after the Brighton and Croydon Companies amalgamated no trace of the locomotives of the latter railway can be found. They do not appear to have names in the Brighton Company's stock, neither can they be traced in the list of SER locomotives, excepting one engine which we can trace to this company. When the locomotives were apportioned at the dissolution, the SER became possessed of an engine named *Man of Kent*; it was built by G. & J. Rennie, London, the cylinders were 15 inch diameter, 18 inch stroke, with 5 foot 6 inch diameter of driving wheels. This locomotive became No. 27 on the SER, but no light can be thrown on the point whether this was a Brighton or a Croydon engine originally. The SER also possessed an engine built by Sharp, Roberts & Co., having 12 inch by 18 inch cylinders, and 5 foot 6 inch driving wheels, numbered and named 45 *Brighton*, afterwards No. 4 on the London & Greenwich Railway.

On 1 February, 1847, S. Kirtley succeeded John Gray as head of the locomotive department, the latter's predecessor until May, 1845, having been a Mr Statham. In December, 1847, John C. Craven was appointed locomotive superintendent.

It may be interesting to give such information as is accessible of the London & Croydon Railway Company's locomotive stock before describing that of the LBSCR:

No. of Engine	Name of Engine	Date	Maker	Maker's No.	Cylinders	Dia. driving wheels
1	Surrey	1838	Sharp, Roberts	4	13 x 18	5 feet 6 inches
3	Sussex	1838	Sharp, Roberts	7	13 x 18	5 feet 6 inches
4	Kent	1838	Sharp, Roberts	19	13 x 18	5 feet 6 inches
5	London	1839	Sharp, Roberts	25	13 x 18	5 feet 6 inches
7	Hercules	1839	Sharp, Roberts	33	14 x 18	5 feet 0 inches
8	Sydenham	1839	Sharp Roberts	43	13 x 18	5 feet 0 inches
2	Croydon	1838	G. & J. Rennie		13 x 18	5 feet 0 inches
6	Archimedes	1839	G. & J. Rennie		13 x 18	5 feet 0 inches
	Coryndon	1840	J. Chanter		13 x 18	5 feet 6 inches
	Achilles	1843	Sharp, Roberts	229	14 x 20	4 feet 6 inches
		1843	Sharp Roberts	234	14 x 18	5 feet 6 inches
		1843	Sharp Roberts	237	14 x 18	5 feet 6 inches
		1843	Sharp Roberts	245	14 x 18	5 feet 6 inches

Engines built to joint orders of L. & C., S.E. and L. & B. Railways:

		Date	Maker	Maker's No.	Cylinders	Dia. driving wheels
		1844	Sharp, Roberts	247-8	14 x 18	5 feet 6 inches
		1844	Sharp, Roberts	267-8	15 x 18	5 feet 6 inches
		1844	Sharp Roberts	270-1	15 x 18	5 feet 6 inches
		1844	Sharp Roberts	275-6	15 x 18	5 feet 6 inches
		1845	Sharp, Roberts	285-6	15 x 20	5 feet 6 inches
		1845	Sharp, Roberts	295-8	15 x 20	5 feet 6 inches

In 1838, Sharp, Roberts & Co., of Manchester, supplied the LCR with three single-wheel engines named *Surrey*, *Sussex*, and *Kent*. These engines had cylinders 13 inch diameter by 18 inch stroke, with driving wheels 5 foot 6 inch diameter, and leading and trailing wheels 3 foot 6 inch diameter. The boilers were 3 foot 4 inch diameter by 8 feet long, and contained 132 tubes of 1⅝ inch diameter, with a heating surface of 470.58 square feet, and the fireboxes were 3 feet by 3 feet 6 inches, with a heating surface of 46.62 square feet. They weighed 12½ tons. In the following year, 1839, they supplied two more of the same class, named *London* and *Sydenham*. This latter engine was afterwards re-named *Victoria*. In the same year they supplied an engine named *Hercules*, having cylinders 14 inch diameter, stroke of pistons 18 inches, diameter of driving wheels 5 feet 0 inches. This was very probably a four-coupled engine.

In 1838 and 1839 respectively G. & J. Rennie & Co., of Blackfriars, constructed two four-wheel coupled goods engines named *Croydon* and *Archimedes*, which were used as the New Cross bank pilots. These locomotives had outside horizontal cylinders, 13 inch diameter by 18 inch stroke, with connecting rods working upon crank pins in the driving wheels; these were 5 foot diameter, and were coupled to the leading wheels, whilst the engines also had a pair of trailing wheels 3 foot 6 inch diameter, and weighed empty 13 tons. The boilers were 3 foot 3⅝ inch diameter by 8 feet in length, containing 117 tubes of 1⅞ inch diameter, and 4 of 1⅝ inches, with a heating surface of 496.3 square feet, and the fireboxes were 3 feet by 3 feet 7 inches, having a heating surface of 60.14 square feet. The driving wheel tyres had no flanges, and the valve gear was of the type known as Carmichael's. This motion had a single fixed eccentric for each valve, and arms at the ends of the eccentric rods which projected on either side to engage with a rocking shaft. This arrangement was first introduced by Carmichael, of Dundee, in 1818, in the twin ferry boats working across the Firth of Tay. It was modified for locomotive engines by M. Cavé, of Paris, in 1835.

In 1840 John Chanter & Co. supplied an engine named *Coryndon*; the cylinders were 13 inch diameter with an 18 inch stroke, the driving wheels being 5 foot 6 inch diameter. This locomotive had a furnace in which coal was fed through a hopper upon an inclined grate, and the gases as they were evolved were met by highly heated air. This arrangement was for the prevention of smoke, and another engine of the same type was subsequently tried on the Liverpool & Manchester Railway, but on both lines the results obtained were inferior to those from the ordinary firebox. This engine, after the dissolution, became No. 79 on the SER.

Sharp, Roberts & Co., in 1843, supplied the company with a four-wheel coupled engine named *Achilles*; **the leading and driving wheels were 4 foot 6 inch diameter, the trailing 3 foot 6 inch diameter.** The cylinders were 14 inches by 20 inches. In the same year they supplied three 5 foot 6 inch singles, having cylinders 14 inches by 18 inches. In 1844 Sharp, Roberts & Co. supplied eight engines of the same class to the order of the joint committee. Two of these engines had 14-inch cylinders, the remainder having 15 inch cylinders. In 1845 this engineering firm supplied six engines of the same class, but with increased dimensions, the cylinders being 15 inches by 20 inches

Below: One of the engines built by Bury, Curtis & Kennedy, 1841–1842. *See Chapter 2.*

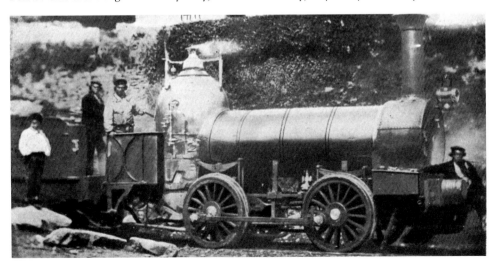

Chapter 2

LBSCR Locomotives, 1841–1849

We come now to the locomotives of the LBSCR, but to give a complete description of all the classes of engines that the company has possessed from its commencement would be almost impossible, because when the joint committee was dissolved some of the Brighton company's engines became the property of the SER, and their history cannot now be traced.

Fortunately, we can trace every engine the Brighton company possessed after the engines were allotted to the two companies, and at the same time we have a record of some engines built before the amalgamation with the Croydon line, which were purely and simply Brighton engines.

It is a remarkable fact that at the end of 1869 this comparatively small company possessed no less than seventy-two different types of locomotives. The number of engines it owned at that date was 233, and if to these were added those previously sold or broken up, the total number of engines built or purchased to that date would have exceeded 270, representing the work of no less than twenty different firms. Perhaps it is stranger still that J. C. Craven, who was locomotive superintendent from 1847 to 1870, should have constructed so many different classes under his own supervision, no more than twelve engines of the same class being built to one set of drawings, and this being only in one case, which order was given to an outside firm. In the early days, however, it must be remembered that private firms generally supplied locomotives to their own standard designs, a fact which would to a large extent account for such a variety of types, but which, nevertheless, adds considerable interest to the locomotive history of the LBSCR.

Down to 1849 the boilers of all the engines had lagging of mahogany, polished, and bound with brass bands, excepting those built by Longridge and by Wilson, which had painted wooden lagging, the strips being red and green alternately. Then, until 1870, those engines that were painted were a Brunswick green, lined with a broad black band having a fine white line on each side, and the frames were crimson lake. This, with their polished brass or copper domes, and safety valve funnels, gave them a very smart and imposing appearance; none of them were named, except a few of the first engines on the line, and two six-wheels coupled engines purchased in 1854, the *Orestes* and *Europa*.

We now append a list of the engines the company had after the dissolution of the joint committee, from which time our history of each engine individually will date. The locomotives will be described in chronological, not numerical order throughout, excepting the engines down to December, 1849. The following is a list of LBSCR locomotives to that date.

LBSCR Locomotive Stock, Nos 1–110:

No.	Maker	Date built or purchased	Class
1	Bury, Curtis & Kennedy	Mar, 1841	Single Passenger
2	Bury, Curtis & Kennedy	Mar, 1841	Single Passenger
3	Bury, Curtis & Kennedy	July, 1841	Single Passenger
4	Bury, Curtis & Kennedy	Oct, 1842	Single Passenger
5	Bury, Curtis & Kennedy	Oct, 1842	Single Passenger
6	Bury, Curtis & Kennedy	Nov, 1842	Single Passenger
7	Bury, Curtis & Kennedy	July, 1841	Single Passenger
8	G. & J. Rennie	Dec, 1841	Single Passenger
9	G. & J. Rennie	Oct, 1840	Single Passenger
10	G. & J. Rennie	Oct. 1840	Single Passenger
11	G. & J. Rennie	Feb, 1843	Single Passenger
12	Fairbairn & Sons	Sept, 1841	Single Passenger
13	Fairbairn & Sons	May, 1842	Single Passenger
14	Fairbairn & Sons	June, 1842	Single Passenger
15	J. G. Bodmer	Dec, 1845	Single Passenger
16	R. & W. Hawthorn	July, 1845	Single Passenger
17	R. & W. Hawthorn	July, 1845	Single Passenger
18	R. & W. Hawthorn	May, 1845	Single Passenger
19	Sharp, Roberts	Mar, 1849	Single Passenger
20	Sharp, Roberts	Feb, 1849	Single Passenger
21	Sharp, Roberts	Feb, 1838	Single Passenger
21	Sharp, Roberts	Feb, 1849	Single Passenger
22	Sharp, Roberts		Single Passenger
23	Sharp, Roberts		Single Passenger
24	Sharp, Roberts		Single Passenger
25	Sharp, Roberts	July, 1839	Single Passenger
26	Sharp, Roberts	Oct, 1844	Single Passenger
27	Sharp, Roberts	Dec, 1844	Single Passenger
28	Sharp, Roberts	April, 1845	Single Passenger
29	Sharp, Roberts	June, 1845	Single Passenger
30	Sharp, Roberts	June, 1845	Single Passenger
31	Sharp, Roberts	June, 1845	Single Passenger
32	Sharp, Roberts	July, 1845	Single Passenger
33	Sharp, Roberts	Feb, 1844	Single Passenger
34	Jones & Potts	July, 1845	Single Passenger
35	Jones & Potts	July, 1845	Single Passenger
36	Jones & Potts	Oct, 1845	Single Passenger
37	Jones & Potts	Jan, 1846	Single Passenger
34	Sharp, Roberts	April, 1849	Single Passenger
35	Sharp, Roberts	Oct, 1849	Single Passenger
36	Sharp, Roberts	Nov, 1849	Single Passenger
37	Sharp, Roberts	Nov, 1849	Single Passenger

No.	Maker	Date built *or purchased*	Class
38	Sharp, Roberts		Single Passenger
39	Sharp, Roberts	Jan, 1849	Single Passenger
40	Sharp, Roberts		Single Passenger
41	Sharp, Roberts		Single Passenger
42	Sharp, Roberts	Oct, 1841	Single Passenger
43	Sharp, Roberts		Single Passenger
44	Sharp, Roberts		Single Passenger
45	Sharp, Roberts	Nov, 1844	Single Passenger
46	Sharp, Roberts		Single Passenger
47	Sharp, Roberts	Mar, 1849	Single Passenger
47	T. Hackworth	June, 1847	Single Passenger
48	T. Hackworth	Sept, 1846	Single Passenger
49	T. Hackworth	July, 1848	Single Passenger
50	T. Hackworth	May, 1848	Single Passenger
51	T. Hackworth	May, 1848	Single Passenger
53	T. Hackworth	Nov, 1846	Single Passenger
54	T. Hackworth	Mar, 1847	Single Passenger
56	T. Hackworth	June, 1847	Single Passenger
57	T. Hackworth	Aug, 1847	Single Passenger
58	T. Hackworth	Sept, 1847	Single Passenger
59	T. Hackworth	Oct, 1847	Single Passenger
60	T. Hackworth	Jan, 1848	Single Passenger
61	E. B. Wilson	June, 1847	Single Passenger
62	E. B. Wilson	July, 1847	Single Passenger
63	E. B. Wilson	July, 1847	Single Passenger
64	E. B. Wilson	July, 1847	Single Passenger
65	E. B. Wilson	Aug, 1847	Single Passenger
66	E. B. Wilson	Aug, 1847	Single Passenger
67	E. B. Wilson	Mar, 1848	Single Passenger
68	E. B. Wilson	April, 1848	Single Passenger
69	E. B. Wilson	Aug, 1848	Single Passenger
70	E. B. Wilson	June, 1847	Single Passenger
71	Sharp, Brothers & Co.	Dec, 1848	4-coup. Passenger
72	Sharp, Brothers & Co.	1848	Single Passenger
73	Sharp, Brothers & Co.	1848	Single Passenger
74	Sharp, Brothers & Co.	1847	Single Passenger
75	Sharp, Brothers & Co.	1848	Single Passenger
76	Sharp, Brothers & Co.	1847	Single Passenger
77	Sharp, Brothers & Co.	1847	Single Passenger
78	Sharp, Brothers & Co.	1847	Single Passenger
79	Sharp, Brothers & Co.	1847	Single Passenger
80	Sharp, Brothers & Co.	1847	Single Passenger
81	Sharp, Brothers & Co.	June, 1847	Single Passenger
82	Sharp, Brothers & Co.	1847	Single Passenger
83	Sharp, Brothers & Co.	1847	Single Passenger
84	Stothert & Slaughter	Mar, 1848	Single Passenger
85	Stothert & Slaughter	Jan, 1848	Single Passenger
86	Stothert & Slaughter	Jan, 1848	Single Passenger
87	Stothert & Slaughter	Jan, 1848	Single Passenger
88	Stothert & Slaughter	Jan, 1848	Single Passenger
89	Stothert & Slaughter	Jan, 1848	Single Passenger
90	Stothert & Slaughter	Sept, 1847	4-coup. Passenger
91	Stothert & Slaughter	Sept, 1847	4-coup. Passenger
92	Stothert & Slaughter	Sept, 1847	Single Passenger
93	Stothert & Slaughter	Oct, 1847	Single Passenger
94	Stothert & Slaughter	Oct, 1847	Single Passenger

No.	Maker	Date built or purchased	Class
95	Stothert & Slaughter	Mar, 1848	Single Passenger
96	Bury, Curtis & Kennedy	Dec, 1843	4-wheel Goods
97	Bury, Curtis & Kennedy	Nov, 1844	4-wheel Goods
98	Bury, Curtis & Kennedy	May, 1845	4-wheel Goods
99	Bury, Curtis & Kennedy	Sept, 1845	4-wheel Goods
96	R. B. Longridge		6-coupled Goods
97	R. B. Longridge	July, 1848	6-coupled Goods
100	R. B. Longridge		6-coupled Goods
101	R. B. Longridge		6-coupled Goods
102	R. B. Longridge		6-coupled Goods
103	R. B. Longridge	1847	6-coupled Goods
104	R. B. Longridge		6-coupled Goods
105	R. B. Longridge		6-coupled Goods
106	Stothert & Slaughter	Jan, 1847	6-coupled Goods
107	Stothert & Slaughter	Dec, 1846	6-coupled Goods
108	Sharp, Brothers & Co.	July, 1847	4-coup. Passenger
109	Sharp, Brothers & Co.	July, 1847	July, 1847 "
110	Sharp, Brothers & Co.		No record

By reference to the foregoing list it will be seen that the first engines that we have to describe are Nos 1–7, built by Bury, Curtis, & Kennedy, Clarence Foundry, Liverpool. It will be seen from *Fig. 1* that they were of the standard Bury type, with bar framing, round firebox, and a single pair of driving wheels 5 feet in diameter, whilst the cylinders were 14 inch diameter by 20 inch stroke. The boilers of these engines were 3 foot 6 inch diameter by 10 feet in length, containing 96 tubes of 2⅛ inch diameter. At a later date a small pair of trailing wheels was added to these engines. No. 1 was sold in 1858.

No. 4 was again altered, being fitted with a pair of trailing wheels, 5 feet in diameter, coupled to the driving wheels, and a new boiler, the tender being removed and a coal bunker and tank substituted, with a footplate running round the side of the engine *(Fig. 2)*. The diameter of the boiler was 3 feet 5 inches, length 10 feet 2¾ inches, diameter of firebox 3 feet 6 inches, leading to driving centres 7 feet 2 inches, driving to trailing 6 feet 6 inches. The mahogany lagging was discontinued, and the boiler clothed with sheet iron.

It once happened that No. 4 over-ran the turntable at London Bridge and fell over

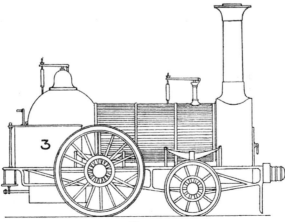

Fig. 1

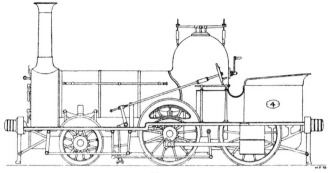

Fig. 2.

the viaduct into Bermondsey Street, wheels uppermost. Not much damaged, it was drawn through the streets and put on the rails again at Bricklayers' Arms Station.

In 1850 No. 6 was reconstructed as a six-wheeled engine, the diameter of the leading and driving wheels being 4 feet 9 inches, and that of the trailing wheels 4 feet 0 inches. A new pair of cylinders, 13 inch diameter by 18 inch stroke, was also put in, and new link motion, axle boxes and framing, and a tank was placed over the boiler to contain 500 gallons of water. No. 7 was a few years later re-numbered No. 5.

Engines Nos 8–11 *(Fig. 3)* were built by G. & J. Rennie, and were of the same type as the *Brighton* and *Shoreham*, previously described. No. 8 was named *Satellite*, and later re-numbered No. 23, No. 9 was named *Eagle* and ran the first train from Brighton to Shoreham, No. 10 was *Vulture*, and No. 11 *Kentish Man*. In the company's books the date of *Eagle* is October, 1840, although it opened the Shoreham line in May of the same year. It then probably belonged to the contractors who made the line. Such engines were often purchased by the railway companies in those days. No. 10 was afterwards converted into a well-tank engine.

On the morning of 17 March, 1853, the boiler of No. 10 exploded at Brighton Terminus, killing the driver and fireman, and also a fitter who was to have accompanied the engine on a trial trip after repairs. The engine was attached to the 7.15 a.m. train to Littlehampton at ten minutes to seven, when the boiler suddenly burst with a terrific report. The barrel was completely destroyed and the tubes scattered, whilst the smokebox fell forward on to the buffer beam, the tank was driven back into a carriage, the piston rods were bent up against the cylinders, and the crank axle was broken into three pieces and the wheels forced off. A portion of the firebox was blown up through the station roof and descended through another part of the roof on to a carriage. This explosion is attributed to the conduct of the driver in subjecting the boiler to too high a pressure. At the subsequent inquiry it was stated that the boiler must have been pressed to nearly 300 lbs per square inch, the working pressure being only 100 lbs. Previous to the explosion this engine had worked the Eastbourne branch.

Engines Nos 12, 13 and 14 were built by W. Fairbairn & Sons, of Manchester, and were of the same class as the preceding, but with cylinders 14 inch diameter by 22 inch stroke, the driving wheels being 5 foot 6 inch diameter *(Fig. 4)*. No. 12 was rebuilt in December, 1858, No. 13 was sent to Deptford Saw Mills in 1854 to be used as a stationary engine.

An engine of this class, piloted by one of Bury's singles, was bringing down a heavy train from London to Brighton, on 2 October, 1841, when the pilot engine ran off the road and buried itself in the side of Copyhold cutting, between the Ouse Valley viaduct and Hayward's Heath Station, two passengers being killed and three coaches smashed. The cause was attributed to a four-wheeled engine preceding a six-wheeled one, and it was in consequence of this that engines 1–4 had a pair of trailing wheels added.

R. & W. Hawthorn & Co., of Newcastle-on-Tyne, built engines Nos 16–19. These had cylinders 15 inch diameter by 21 inch stroke, and driving wheels 5 foot 6 inch diameter. No. 19 was re-numbered 15, and with Nos 16 and 17 were converted into tank engines, Nos 15 and 17 being eventually sold to the East Kent Railway. No. 16, in June, 1854, was sold to the Amsterdam Water Works, in Holland; and No. 18 was converted into a four-coupled tank engine.

In December, 1845, J. G. Bodmer, of Manchester, supplied the company with a single-wheel express engine, No. 15, the driving wheels being 6 feet in diameter *(Fig. 5)*. The cylinders were of a special design, each having two pistons, moving simultaneously in opposite directions, the reciprocating parts attached to the one counteracting the momentum of those attached to the other. They were 15 inch diameter by 20 inch stroke, each piston having a travel of 10 inches. The firebox was of the Bury type. This engine was re-numbered 7 in Craven's time, and ran the first 5 p.m. train from Brighton to London. The patent piston arrangement not proving a success, the engine was fitted

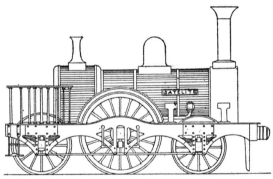

Fig. 3.

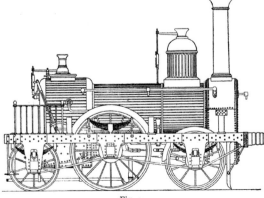

Fig. 4.

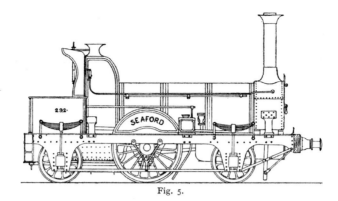

Fig. 5.

with ordinary cylinders in January, 1859, and then continued to do good work for some years. After the appointment of Stroudley as locomotive superintendent, No. 7 was named *Seaford*, and re-numbered 292, and again 367. An engine of this class, but with cylinders 16 inches by 30 inches, was taken over by the SER when the Joint Locomotive Committee was dissolved, and re-numbered 123.

Sharp, Roberts & Co., Atlas Works, Manchester, were the builders of engines Nos 20–33, and as will be seen by reference to the list, their dates ranged from 1838 to 1849; no doubt some of the Croydon engines are included among them. These were known as 'Sharp's small singles' *(Fig. 6)*. Their chief dimensions were: Cylinders 15 inch diameter by 20 inch stroke, distance between centres 2 feet 5 inches, diameter of leading and trailing wheels 3 feet 6 inches, diameter of driving wheels 5 feet 6 inches, leading to driving centres 6 feet 3 inches, driving to trailing 6 feet 11 inches, boiler barrel 3 foot 6 inch diameter by 10 feet long, 139 tubes, outside diameter 2 inches, length 10 feet 6 inches, length of outside firebox 3 feet 9 inches, heating surface of tubes 729 sq. ft, and of firebox 67 sq. ft, weight of engine in working order 22 tons 3 cwt. The boiler had circumferential and longitudinal butt joints, and was made in three rings, the dome being placed on the front ring close to the chimney, giving the engine a humpbacked appearance. One spring balance safety valve was fixed on the dome, and another over the firebox. These engines had outside axle boxes to the leading and trailing wheels, and both inside and outside boxes to the driving wheels; they were

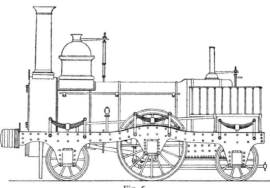

Fig. 6.

fitted with Stephenson's link motion and reversed with a lever.

The next engines were from Jones & Potts, of Newton-le-Willows, Lancashire, and had outside cylinders and single driving wheels without flanges, and were numbered from 34 to 37. The cylinders were 15 inch diameter by 22 inch stroke, and the diameter of the driving wheels 5 feet 11 inches. These engines were constructed on Stephenson's 'long-boiler' patent, having all the wheels in front of the firebox. The latter was of the high-top conical type, with flat sides tapered towards the top.

These engines proved very unsteady when running at any fair speed, owing to the oscillation caused by the outside cylinders and the great overhanging weight behind the trailing wheels, which, together with their flangeless driving wheels, resulted in two engines of this class leaving the road when running with trains. The first of these occurrences was in the autumn of 1846, when a train-load of hops, which was proceeding from Hastings to Brighton, somehow or other left the road just beyond Southerham Junction, near Lewes, and ran into the River Ouse. The other was on 31 May, 1847, on which occasion a passenger train from Brighton to Havant was running through Nutbourne Copse, near Emsworth, at between 30 and 35 miles per hour, when the engine suddenly ran off the rails on the right-hand side, and crossing the Up line at an angle of about 30 degrees, fell over an embankment and plunged, wheels uppermost, into a ditch. The coupling between engine and tender broke, and probably averted a more serious disaster. Kirtley, the locomotive superintendent, had ridden on the engine as far as Bosham, when he alighted and rejoined his wife in a first-class carriage.

These engines were afterwards re-numbered 115, 96, 97 and 98 respectively. No. 96 was made into an inside cylinder four-coupled tender engine, No. 97 into an outside cylinder single-wheeled tank engine, and No. 115 into an inside cylinder four-coupled tender engine, the last-named being at a later date converted into a side-tank engine.

Fig. 7 shows No. 97 after being altered, and will give a good idea of this class when in the old state; it will be noticed that the trailing wheels have been placed behind the firebox, and the boiler shortened by having the first ring of the barrel taken away. The principal dimensions were: Cylinders 15 inch diameter by 22 inch stroke, diameter of leading and trailing wheels 3 feet 10 inches, diameter of driving wheels 5 feet 11 inches, leading to driving centres 6 feet, driving to trailing 7 feet 6 inches, diameter of

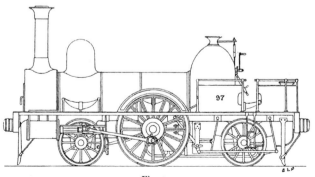

Fig. 7.

boiler 3 feet 6 inches, length 11 feet, height of centre line 5 feet 11 inches, number of tubes 143, outside diameter 2 inches, length 11 feet 5 inches, weight 26 tons 10 cwt, of which 10 tons were on the driving wheels. No. 97 was altered thus in October, 1852.

Engine No. 115, in *Fig. 8*, the cylinders being 15 inch diameter by 20 inch stroke, diameter of leading and driving wheels 5 feet, and of trailing wheels 3 feet 6 inches, wheel-base 15 feet equally divided, boiler 3 feet 3 inches in diameter by 11 feet 2 inches in length, containing 114 tubes of 2 inch outside diameter, and 11 feet 6¾ inches long. The firebox was 3 feet 8 inches in length, and the heating surface was, of tubes, 704 sq. ft, and of firebox 66 sq. ft, the weight of engine being 27 tons 10 cwt. The side tanks held 420 gallons of water, and the tank under the footplate 231 gallons. No. 115 was again re-numbered 260 and finally 404, and scrapped in Feb., 1879.

The next class of locomotive to be dealt with is a series of single driving wheel

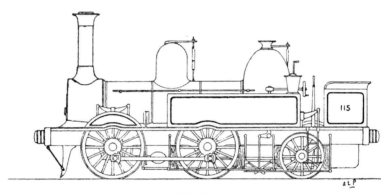

Fig. 8.

express engines by Sharp, Brothers & Co., as the firm had then become. These engines have always been known on the line as the 'Big Sharps', and in their time did some very good work, both at running and hauling, for the loads between London and Brighton were always heavy. These locomotives were sixteen in number, all of the same class, except one – No. 19 – which was of the smaller type. Their numbers, makers' numbers and dates were as follows:

Engine No.	Maker's No.	Date	Engine No.	Maker's No.	Date
34	573	April, 1849	42		
35	609	Oct, 1849	43	558	Jan, 1849
36	610	Nov, 1849	44		
37	613	Nov, 1849	45	559	Jan, 1849
38			46		
39	557	Jan, 1849	47		March, 1849
40			19	571	March, 1849
41			21	568	Feb, 1849

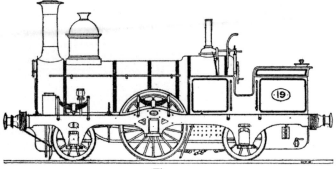

Fig. 9.

The small single, No. 19, was very similar in dimensions to the first lot of Sharp's singles, No. 27 class. This locomotive, however, was converted into a tank engine in 1860, *Fig. 9.* From this it will be seen that the wheels were not altered in any way, but merely a tank substituted for the tender. The cylinders were 15 inch diameter by 20 inch stroke, diameter of leading and trailing wheels 3 feet 6 inches, diameter of driving wheels 5 feet 6 inches, distance between leading and driving centres 6 feet 6 inches, and between driving and trailing 7 feet 6 inches. The boiler was 9 feet 9½ inches long by 3 foot 10¾ inch external diameter, the height of its centre line being 5 feet 11½ inches, and the length of the outside firebox 3 feet 11 inches. A small tank for water was fixed under the cylinders in front of the leading wheels, and another under the foot-plate behind the trailing wheels. The brake arrangement was curious, being fitted on the right-hand side of the engine only, one block on the driving and one on the trailing wheel. The total length of the engine over buffers was 26 feet 9½ inches, and the total weight 28 tons 8 cwt, there being 8 tons 18 cwt on the leading wheels, 10 tons 18 cwt on the driving, and 8 tons 12 cwt on the trailing.

Fig. 10 illustrates the large Sharp singles, which had cylinders 15 inch diameter by 22 inch stroke, centres 3 feet apart, diameter of blast pipe orifice 5¼ inches, diameter of leading and trailing wheels 3 feet 8 inches, and of driving wheels 6 feet 6 inches, these being the largest in diameter of any locomotives on the line up to that time, and for some years after. The distance between leading and driving centres was 7 feet 0½

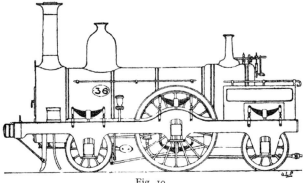

Fig. 10.

inches, and between driving and trailing 7 feet 3 inches. The boiler was made in four rings, with the steam dome on the second, and was 11 feet long by 3 feet 10⅞ inch external diameter, containing 162 brass tubes, 2 inch diameter by 11 feet 5 inches long. The length of the outside firebox was 4 feet 1½ inches, breadth 4 feet 3½ inches, and depth below centre line of boiler 4 feet 2½ inches, the heating surface of tubes being 861.2 sq. ft, and of firebox 76 sq. ft. There was a lock-up safety valve on the steam dome, and two spring balance safety valves over the firebox. The outside frames were of the sandwich pattern, and the inside frames were plates running from the cylinders to the front of the firebox. The latter were not placed in the usual position, but were fixed between the cranks and eccentrics, the driving wheels having bearings in these frames, to which also the pumps were fixed. The regulator was of the revolving disc type. These engines were very light, only weighing 23 tons, of which 7½ tons were on the leading wheels, 10½ tons on the driving wheels and 5 tons on the trailing wheels.

In July, 1877, No. 37 was re-numbered 360 and it was scrapped in March, 1882; in July, 1877, No. 39 was re-numbered 361.

In 1846 John Gray, the locomotive superintendent, prepared designs for a class of express engine of the single driving-wheel type, and comprising some very novel features. These locomotives had an expansive valve gear, known as the 'horseleg motion', patented by Gray, who employed a species of link, vibrating upon an axis at its lower end, and by an elaborate arrangement of reversing gear, the end of either eccentric rod could be engaged in its corresponding curved groove, there being two grooves on opposite sides of the link. The eccentric rods, moreover, could be adjusted to give their motion to an arm at variable distances from the axis upon which it vibrated, which arrangement varied the travel of the valve, and also the lead at the same time, and where there was already considerable lap, the variation of lead proceeded more rapidly than the variation in the travel of the valve. T. Hackworth & Co., Soho Ironworks, Shildon, had the order for constructing these twelve engines, numbered 47–51, 53, 54, 56–60. The dates of these engines will be seen by reference to the previous list. No. 49 is shown in *Fig. 11*, the principal dimensions being – diameter of leading and trailing wheels 4 feet 0 inches, diameter of driving wheels 6 feet 0 inches, cylinders 15 inch diameter by 24 inch stroke, boiler 10 feet long by 4 foot 1 inch diameter, length of firebox 3 feet 4 inches. The wheelbase was 14 feet, being divided as follows: Leading

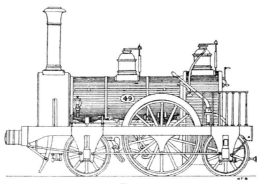

Fig. 11.

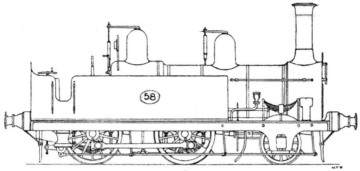

Fig. 12.

to driving 7 feet 3 inches, driving to trailing 6 feet 9 inches. Length of firebox 4 feet 2 inches, centre of boiler from rails 6 feet 3½ inches. The heating surface of the tubes was 700 sq. ft, and that of the firebox 79 sq. ft, giving a total of 779 sq. ft, the working pressure being 100 lbs per sq. in. As seen by the illustration, there were two steam domes, both alike, with square seatings and spring balance safety valves. Some of the engines, however, had only one dome.

These engines proved very good for fast running, easily attaining a speed of 40 miles per hour with a fairly heavy load. They had inside boxes for the driving wheels and outside for the leading and trailing, and were thus the forerunners of the well-known 'Jenny Lind' class, which we shall describe later. Before doing so, we will notice some of the very radical metamorphoses these engines went through during their existence.

Nos 56 and 58 were converted experimentally to Crampton's principle, and resembled in external appearance the 'Folkestone' class on the SER, there being no wheels on the crank axle, which was connected to the trailing wheels by coupling rods. The results obtained from these two engines proving unsatisfactory, they were again altered, together with engines Nos 49–52, into four-wheels coupled goods engines, Nos 49–51 having 16 inch cylinders, and Nos 52, 56 and 58 had 15 inch cylinders, the stroke being 24 inches and the diameter of the driving and trailing wheels 4 feet 9 inches in all six engines. Subsequently Nos 49 to 51 were further altered into goods tanks, having a pair of leading wheels the same diameter as the driving and trailing and coupled to them, thus becoming six wheels coupled. The wheelbase was 14 feet equally divided, the boiler 10 feet long by 4 foot diameter, containing 145 tubes of 2¼ inch diameter, and the length of the firebox 4 feet 8 inches. These engines weighed 31 tons 14 cwt, having 10 tons 6 cwt on the leading wheels, 12 tons 14 cwt on the driving wheels, and 8 tons 14 cwt on the trailing wheels. No. 58 was converted into a four-wheels coupled tank engine, shown in *Fig. 12*, the chief dimensions being: diameter of leading wheels 3 feet 4 inches, diameter of driving and trailing wheels 4 feet 9 inches, wheel-base from leading to driving 7 feet 9¾ inches, and from driving to trailing 7 feet 5½ inches, cylinders 16 inch diameter by 24 inch stroke, diameter of boiler 4 feet, 145 tubes of 1⅞ inch diameter, length of firebox 4 feet 1½ inches, and capacity of tanks 593 gallons.

Engines Nos 53–55, 57, 59 and 60 were converted into four-wheels coupled

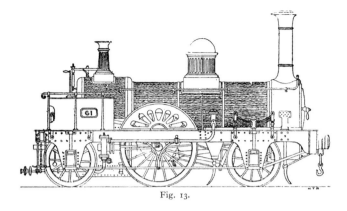

Fig. 13.

passenger engines, having cylinders 15 inches by 24 inches, and 5 foot 6 inch coupled wheels.

In May, 1847, No. 54 wras standing a little way from the Brighton station, the driver and fireman being off the engine at the time, when it suddenly started off, unattended, and dashing into the station collided with a luggage van, which it literally lifted off the rails on to the platform, sending it with considerable force into the parcels office and breaking down the wall. The large buffer springs and posts at the end of the rails were likewise shattered, but, strange to say, the engine did not leave the rails.

Engine No. 47 was, shortly after the dissolution of the Joint Committee, re-numbered 55, and No. 48 was re-numbered 52 in May, 1862. They were again re-numbered 114 and 110 respectively.

We next come to the well-known 'Jenny Lind' class, which, as stated above, were very similar to the 49 class in detail and appearance, excepting, of course, the valve motion. E. B. Wilson & Co., of Leeds, constructed ten of these engines for the Brighton Co., which were numbered from 60–69, but the number of the first of these had to be changed from 60 to 70, as the last of the Hackworth engines, delivered in the beginning of 1848, was through an oversight also numbered 60. The Jenny Lind type, which was extensively adopted on all the principal British railways, was designed specially for the Brighton Railway by the late David Joy, of valve gear fame, to whom we are indebted for the particulars of these engines. The first engine, No. 60, was named *Jenny Lind*. No. 69 engine was slightly different in dimensions from the rest of the class, but practically similar in detail.

Fig. 13 shows an engine of this class according to the original design. The boiler was lagged with polished mahogany, but afterwards the lagging was painted with alternate strips of green and red, and finally the boilers were clothed with sheet iron and painted the standard Brunswick green colour. The principal dimensions of Nos 61–68 and 70 were: Cylinders 15 inch diameter by 20 inch stroke, diameter of leading and trailing wheels 4 feet, of driving wheels 6 feet, wheel-base from leading to driving 7 feet, and from driving to trailing 6 feet 6 inches, making a total of 13 feet 6 inches. The centre line of the boiler was 5 feet 9 inches from the rail. The boiler contained 120 tubes 2 inch diameter, giving 720 sq. ft of heating surface, the firebox having 80 sq. ft more, giving a total of 800 sq. ft. The weight in working order was 24 tons 1 cwt. The

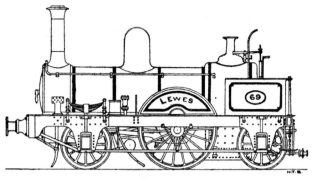

Fig. 14.

boiler was fed by a pump fixed to the outside framing between the driving and trailing wheels, and actuated by a connecting rod and crank pin fixed in the outside boss of the driving wheels. In order to obtain from the firebox sufficient heating surface and at the same time to keep its length to a minimum, it was made as wide as possible, the inside frames finishing at the front or throat-plate of the firebox, the latter being flanged outside to overlap the framing. This arrangement was frequently adopted on Brighton engines. The slide valves were worked by Stephenson's link motion, suspended on one side only. In November, 1861, No. 61 was rebuilt. Nos 62 and 63 were afterwards converted into four-coupled tank engines, which we shall describe later.

As previously mentioned, engine No. 69 was slightly different from the previous engines of this class. This engine was known as the 'Jenny'. The diameter of the driving wheels was increased to 6 feet 3 inches, or 3 inches larger than the other nine engines. This engine also had a mid-feather or double firebox and was much bigger, having an increased heating surface of 1,334 sq. ft as against 800 sq. ft in the others. The pumps for feeding the boiler were placed inside or between the frames of the engine, and were worked by the ram being connected to the engine crosshead. They had previously been placed outside the frames, but in this position they were too much exposed to the atmosphere, and in the winter months they were frequently frozen up. Another difference in this engine was the polished brass chimney cap, whilst the other engines had a copper one. The cylinders were 16 inch diameter, stroke of pistons 22 inches, the wheel-base being 14 feet 6 inches, equally divided. The centre of the boiler was 6 feet 3¼ inches from the rail level, and the weight 27 tons 15 cwt.

Fig. 14 shows this engine after the mahogany lagging was taken off and sheet iron substituted, and may be taken to represent all this class after this alteration was made, with the exception of the driving splasher. It will also be noticed that the fluted dome and safety valve casings had been dispensed with, and sheet brass casings put in their place. No. 69 was named *Lewes* in Stroudley's term of office, and was afterwards re-numbered 295, and then again altered to 365, these changes taking effect when new engines were constructed with the numbers 69 and 295 respectively.

As previously stated, engines Nos 62 and 63 were altered by Craven into passenger tank engines, one of which, No. 63, is shown in *Fig. 15*, the principal dimensions of these two engines being: Cylinders 15 inch diameter by 20 inch stroke, diameter of

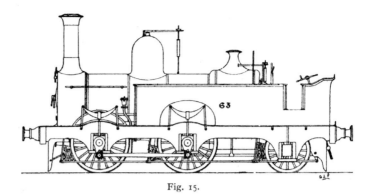

Fig. 15.

leading and driving wheels 5 feet, diameter of trailing wheels 3 feet 9 inches, from leading to driving centres 7 feet 6 inches, and from driving to trailing centres 8 feet 6 inches, height of centre line of boiler from rails 6 feet 3 inches. The bunker was capable of carrying 1 ton 5 cwt of coke, whilst the side tanks would hold 539 gallons of water, and the well tank under the footplate 254 gallons. One spring balance safety valve was fixed on the steam dome, and another was placed directly over the firebox. The steam regulator was of the equilibrium or double beat type, the original regulator valve used in these engines being the ordinary vertical slide valve. Their weight was 35 tons 10 cwt, being distributed as follows – 11 tons 16 cwt on the leading wheels, 12 tons 12 cwt on the driving wheels, and 11 tons 2 cwt on the trailing wheels. In October, 1875, Nos 62 and 63 were re-numbered 301 and 302; in December, 1876, No. 301 was re-numbered 351, and in February, 1879, it was broken up. In December, 1877, 302 was re-numbered 402, and in June, 1880, altered to 501, and broken up in June, 1882.

The next engines which we have to deal with are thirteen built by Sharp, Brothers & Co., and numbered 71 to 83. Except No. 71, which was four coupled, these engines were of the standard 5 foot 6 inch single driving wheel type of that firm.

Engine No. 71 had the driving and trailing wheels coupled, the diameter being 6 feet. The cylinders were 15 inch diameter by 20 inch stroke. Except that this engine was rebuilt in June, 1861, no further particulars can be obtained.

To return to the singles, the principal dimensions were as follows: Cylinders 15 inches by 20 inches, diameter of leading and trailing wheels 3 feet 6 inches, diameter of driving wheels 5 feet 6 inches, of boiler 3 feet 6 inches, length 10 feet, containing 139 tubes 2 inch diameter by 10 feet 6 inches long; height of centre line from rail level 5 feet 11 inches, length of firebox 3 feet 9 inches. As was the firm's practice, the steam dome was placed on the first ring of the barrel, with one spring balance safety valve thereon, and another over the firebox. The tubes contained 729 sq. ft of heating surface, and the firebox 67 sq. ft, making a total of 796 sq. ft. These engines had 8 tons 2 cwt on the leading wheels, 8 tons 18 cwt on the driving wheels, and 5 tons 16 cwt on the trailing wheels, making a total of 22 tons 16 cwt. From the foregoing particulars it will be seen that this class of engine, though rather small, were capable of, and during their time did very good work with loads suitable for their cylinders and weight. Their numbers, makers' numbers and dates were:

Engine No.	Maker's No.	Date	Engine No.	Maker's No.	Date
71		1848	78	443	1847
72	457	1848	79	446	1847
73	469	1848	80	410	1847
74	442	1847	81	424	1847
75	470	1848	82	427	1847
76	456	1847	83	433	1847
77	440	1847			

An engine of this class, No. 72, was converted into a four-coupled tank by Craven for working the Epsom Downs branch, reference to which will be made later on, and in 1859 another, No. 77, was converted into a single-wheel tank engine.

A curious accident happened to engine No. 82 in the summer of 1851. In order to increase the locomotive facilities between Brighton and the towns of Shoreham, Worthing, Hayward's Heath and Lewes without delaying the through traffic, the company adopted the practice of running 'short' trains on a roundabout system. Arriving at Brighton from Worthing, the engine would run round the train and leave in a few minutes for Hayward's Heath, and on its return to Brighton would leave again for Lewes. Just before noon on Friday, 6 June, 1851, the train arrived at Brighton from Hayward's Heath, and at 12.30 p.m. left again for Lewes. It consisted of four vehicles, a second-class coach, a third-class, a first-class, and lastly a second-class brake (containing two passengers and the guard), the engine running tender first. After passing Falmer there is the steep gradient falling towards Lewes and it was usual to descend this incline with steam shut off, and the hand-brake applied, there being, of course, no continuous brake in those days. About half-way down this incline the line is carried by a brick arch, known as Newmarket Arch, over a bridle or occupation road, and when the train was within about 80 yards of this the tender and engine suddenly left the road, and bearing slightly to the right passed over the Up-line to the south side of the railway, and, forcing down the parapet wall of the arch, fell with a tremendous crash into the road beneath, a depth of 25 feet, dragging after them the second and third-class coaches. The first-class was flung round so that its end was close to the arch, and the remaining brake carriage was thus brought up. The engine and tender were smashed up, the frames being twisted out of all shape, and the two coaches shivered to pieces. Three of the passengers and the fireman were killed on the spot, the driver succumbed to his injuries three days after. This accident is attributed to a shepherd boy having wilfully placed a sleeper across the outer rail of the Down line.

On 27 November, 1851, No. 81, whilst running a passenger train, collided with a cattle train at the drawbridge over the River Arun, near Ford Station, and went over the embankment, dragging the carriages after it. Two passengers and the fireman of No. 81 were injured, the latter so severely that he died a few days afterwards. The driver of the cattle train jumped into the river to save himself, and the driver of No. 81, to whose negligence in disregarding the signals the accident was due, immediately afterwards attempted to commit suicide by cutting his throat; but failing to accomplish this, he leapt into the river, out of which he was dragged by the guard.

Another of this class, No. 79, ran away from Petworth to Horsham on 22 October, 1859. It appears that at 3 a.m. on the above date, the fire was put in the engine, which was in the shed at Petworth, and by 5 o'clock the pressure gauge showed 15 lbs of steam. The cleaner who was in attendance upon the engine at the time wanted to have it moved to enable him to clean parts of the engine which were inaccessible in the position it then occupied, and he left it to ask the fireman, who was resting in a hut close by, to do this. On returning to the running shed, he heard the beat of an engine, and thereupon went back to inform the fireman that he need not come, as the supposed new arrival would pull No. 79 out. Both returned to the shed, and finding to their astonishment that No. 79 was gone, they ran down the yard, and catching sight of the steam from the engine, they gave chase and very nearly succeeded in capturing the runaway, which was not travelling at a high speed, the cleaner actually getting hold of the buffer; but at that moment, falling from exhaustion, neither of them managed to mount the footplate. The engine continued to gain speed, and ran to Horsham, a distance of 17½ miles, before being brought to a standstill. During this journey it smashed down three pairs of gates at level crossings, and a cleaner from the Horsham shed, who was proceeding down the line, noticing the engine coming along with portions of the broken gates on the buffer beam, and observing there was no one in charge, ran after it for about half-a-mile, to a point where the line begins to rise, at which place the engine, not having a large supply of steam, considerably reduced its speed. This enabled him to mount the footplate, reverse the engine, and bring it back to the Horsham shed. The cause of this occurrence was never found out, but when the engine was caught at Horsham, the regulator was found to be wide open.

It will be remembered that that No. 72 was converted into a four-coupled tank engine for the branch train service between Sutton and Epsom Downs, and we now give some particulars of this engine after its conversion *(Fig. 16)*. The diameter of the driving wheels, coupled to the trailing, was 5 feet 6 inches, the cylinder dimensions being as in the original design, 15 inches by 20 inches. The wheelbase was increased to 14 feet, being 6 feet 6 inches from leading to driving centres, and 7 feet 6 inches from driving to trailing. Total length over buffers was 27 feet 2½ inches. This engine was the only one on the line which had the side tanks connected by a saddle tank. This latter rested on the back part of the boiler, and the three tanks combined contained 390 gallons of water; whilst another was fixed under the footplate, behind the trailing wheels, capable of holding 194 gallons, making a total water capacity of 584 gallons.

No. 78 was latterly re-numbered 279, and in December, 1879, altered to 408. No. 81

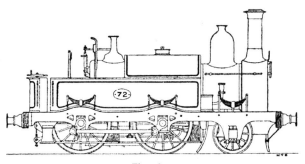

Fig. 16.

was re-numbered 282, and in September, 1879, 407, and was broken up in May, 1880. No. 83 was re-numbered 284, and in September, 1879, 406, and again in June, 1880, 502. Finally it was broken up in March, 1882.

During the same period that Sharp, Brothers and Co. were building engines for the Brighton line, Stothert, Slaughter & Co., of Bristol, were also under orders to supply it with locomotives, rendered necessary by the rapid increase of traffic and the opening of new lines. Two classes of engines were built by this firm, four-coupled and single, the latter, in external appearance, except for the position of the steam dome, bearing a striking resemblance to the GWRs locos, probably due to the fact that Stothert & Slaughter's works were situated on the GWR, and that these engines were designed by the makers, and not by the locomotive superintendent. The engines comprised in this order were two four wheels coupled tender engines numbered 90 and 91, and ten single-wheel expresses, numbered 84–89 and 92–95. Like many of the previous engines, they were not numbered in the same order as they were constructed, as will be seen by the previous list. Engines Nos 90 and 91 had driving and trailing wheels coupled, 5 feet 6 inches in diameter, the cylinders being 15 inch diameter by 22 inch stroke. The boilers had high-topped conical fireboxes, and the engines weighed 28 tons 1 cwt, distributed as follows: 8 tons 6 cwt on the leading wheels, 11 tons 12 cwt on the driving, and 8 tons 3 cwt on the trailing. No. 90 was rebuilt in July, 1857, and 91 also in that year. No further record of these two engines appears to have been retained.

The single wheelers, one of which, No. 89, is illustrated by *Fig. 17*, had the following principal dimensions: Cylinders 15 inch diameter by 22 inch stroke; distance apart of centres 2 feet 7 inches; diameter of driving wheels 6 feet, and of leading and trailing wheels 3 feet 8 inches, centre of leading wheels to centre of driving 7 feet 3 inches, and driving to trailing 6 feet 9 inches. The boilers were constructed in three rings, with the steam dome on the first, the external diameter being 4 feet 1 inch and the length 10 feet. They contained 187 brass tubes, the centre line being 6 feet 5 inches from the rails – a somewhat remarkable dimension in those days, when the chief aim of locomotive designers was to obtain as low a boiler centre as possible, and the more conspicuous in this case as the driving wheels were not particularly large. The outside firebox was 4 feet 2 inches long by 4 feet 1 inch wide, and the inside firebox had a midfeather running, not lengthwise, but from one side to the other. Steam was admitted to the cylinders by a revolving disc valve regulator. The frames of these engines were of the

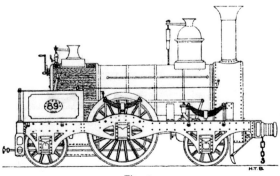

Fig. 17.

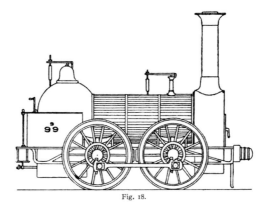

Fig. 18.

sandwich pattern, that is, two plates of iron with wood between. There were four frames in all, two outside and two inside, but the inside frames commenced at the back of the cylinders and finished at the front of the firebox, whilst the outside frames, of course, ran the whole length of the engine. The latter were 4 inches thick, made up of two ½ inch iron plates with a 3 inch layer of oak between, whilst the inside frames were only 2 inches, there being only 1 inch thickness of oak. The wheels had cast-iron bosses, with wrought-iron rims, tyres and spokes, the latter being of T section. These engines weighed 28 tons 8 cwt, distributed as follows – 8 tons 12 cwt on the leading wheels, 10 tons 19 cwt on the driving wheels, and 8 tons 17 cwt on the trailing.

The last engines on the list to describe are four built by Bury, Curtis, & Kennedy. These locomotives were of that firm's standard goods type, had only four wheels and were coupled *(Fig. 18)*, the cylinders being 14 inch diameter by 20 inch stroke. They were employed on the goods and passenger service. Their original numbers we are not acquainted with, but they were re-numbered 96 to 99, and No. 99 when a new engine with this number was built in 1862, was again re-numbered 9⁹9, a small figure 9 being placed above the original number, and after a rebuild with larger boiler and probably a pair of trailing wheels, it ran the Newhaven boat Express, and finished up with the company as pilot at Deptford Wharf. It was then sold to a railway contractor.

It might here be mentioned that this locomotive took part in the dispute between the Brighton and South-Western Companies, which occurred at Havant in December, 1858. No. 98 was re-numbered 100 early in Craven's time. During his regime these engines had cylinders 15 inches by 24 inches, and larger boilers with increased heating surface put in.

Following these engines we find eight goods locomotives, all six-wheels coupled and with tenders, which were purchased from R. B. Longridge & Co., of Bedlington Iron Works, Northumberland. They were numbered 96, 97, and 100–105. Nos 101 and 103, however, were different in construction, being 'long boilers' with high-top fireboxes, while Nos 96, 97, 100, 102, 104 and 105 had the ordinary type of boiler. Nos 101 and 103, *Fig. 19*, also differed from the others of this class in having wheels of cast iron, the spokes of which were of H section, and the rims of T section, the tyres being wrought iron. Their more important dimensions were as follows: Cylinders 15 inch diameter by 24 inch stroke, diameter of wheels 4 feet 9 inches, centre of leading to centre of driving wheels 6 feet 5 inches, driving to trailing 5 feet; boiler 3 foot 6 inch diameter by 13 feet long, containing 111 brass tubes of 2 inch diameter and 13 feet 5 inches long, length of firebox casing 4 feet 2½ inches. The boiler was made in three rings, butt jointed, its centre line being 6 feet 0½ inches from the rail level. It will be seen from the illustration that the frames of the engines were placed inside, or between the wheels.

The following are the principal dimensions of the other six engines: Cylinders 16 inch diameter by 24 inch stroke, diameter of wheels 4 feet 9 inches, centre of leading to centre of driving wheels 6 feet 5½ inches, and driving to trailing 7 feet 7 inches. The boilers were made in three rings, with a steam dome placed on the second one, the barrel being 3 foot 8⅞ inch diameter and 10 feet 6 inches long, and containing 155 tubes of 2 inch diameter, the height of the centre line from the rails being 6 feet 4 inches. The length of the firebox was 4 feet 6½ inches. The regulator was not placed in the dome, but in the boiler at the smokebox end, and fixed in a horizontal position. This regulator consisted of a cylindrical valve, having a seating at the end which was opened and closed by means of a curved guide for the regulator handle, causing the rod to travel longitudinally, and with it the valve, steam being admitted into the regulator cylinder by means of a vertical pipe in the dome. The wheels and motion of some of these engines built by Longridge were afterwards used in certain other engines built at a later period, to which we shall refer later on. Very soon after these engines were supplied to the company, Nos 96, 97 and 100 were numbered 106–108 respectively.

On Monday, 3 October, 1859, the firebox of No. 108 engine exploded under the following circumstances. A heavy goods train, consisting of forty-five loaded waggons, left Lewes just after 9 p.m., drawn by No. 108, and assisted in the rear by No. 50, the heavy bank from Lewes to Falmer – which is on a rising gradient of 1 in 80 – necessitating a pilot when, as in this instance, the train is exceptionally heavy. Between Kingston tunnel and Ashcombe the firebox of the engine mentioned blew up, without warning. The driver and fireman were both thrown with terrific force to a considerable distance ahead, the ashpan and firebars were forced off the engine and the firebox roof stay bars bent out of shape. The explosion acted with a downward force, the rails being bent down where it took place. No particular reason could be given for the explosion, for the engine was in good condition and had run up to this date 236,978 miles, while the boiler pressure was not excessive, being only 90 lbs per sq. in. Next to the engine was a wagon loaded with hops, and behind that a cattle wagon containing two cows.

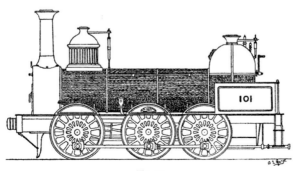

Fig. 19.

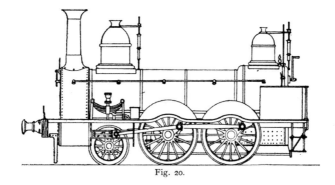

Fig. 20.

This wagon was forced over the truck of hops, and on to the leading engine, by the propulsion of the engine in the rear. When the wagon was once more on *terra firma* the cows were taken out, none the worse for their adventure, and quietly commenced to graze on the embankment. It may be added that the driver and fireman both sustained frightful injuries, from which the former did not recover.

Following these locos came two six-wheels coupled goods engines of the 'long-boiler' type, built by Stothert, Slaughter & Co., and numbered 106 and 107. For some reason the latter engine was the first delivered, in December, 1846, No. 106 following in January, 1847. The important dimensions were: Cylinders, 15 inch diameter by 24 inch stroke, diameter of coupled wheels 4 feet 9 inches, diameter of boiler 3 feet 11 inches, length 13 feet, constructed in four rings, with the steam dome placed on the second one, and containing 163 brass tubes of 1¾ inch diameter. Length of the outside firebox was 4 feet 4 inches, and its top was flush with the barrel of the boiler, a somewhat unusual design in those days when the prevailing style was to have a raised firebox. No. 106 was rebuilt in June, 1855. These two engines were re-numbered 112 and 113 soon after they were built, to make room for the Longridge goods engines, and, as mentioned, they in turn were re-numbered to make room for other engines. In February, 1876, No. 113 was re-numbered 400, and it was broken up in 1878.

In July, 1847, the company purchased from Sharp, Roberts & Co. two four-coupled tender engines, *Fig. 20*. These engines had 'long boilers', and two steam domes, one close up to the chimney and the other over the firebox. The cylinders, which were outside, were 15 inch diameter by 22 inch stroke, the diameter of the driving and trailing wheels being 4 feet 9 inches. Unfortunately, no further particulars of these two locos are available, as no record appears to have been retained. They were numbered 108 and 109, the former being rebuilt in 1854, and No. 111 in March, 1860.

Like the previous engines just mentioned, No. 108 was re-numbered 111 to allow No. 108, Longridge goods engine, to fall into that number. The maker's numbers of these two engines were 430 and 429 respectively. There was also another locomotive by Sharp, Roberts & Co., No. 110, but we have no particulars of it.

With these we close the first ten years of locomotive construction up to the close of 1849 – it will be seen that all the locos, except those altered or converted, had been built by private firms. After that date the company constructed its own stock, although outside firms secured some orders at times, especially between 1864 and 1869.

Chapter 3

John Chester Craven, 1852–1869

The first locomotive built at Brighton Works was a single-wheel tank engine, which, we believe, was designed for working the traffic on the Eastbourne branch. This engine was turned out in May, 1852, and numbered 14. In June of that year another engine of similar design and dimensions was built, and was numbered 26. An illustration of No. 14 is shown in *Fig. 21* the principal dimensions being:Cylinders 13 inch diameter by 20 inch stroke, diameter of leading and trailing wheels 3 feet 8 inches, diameter of driving wheels 5 feet 7 inches, from leading to driving centres 6 feet 6 inches, and from driving to trailing 7 feet 3 inches, making a total wheel-base of 13 feet 9 inches; boiler barrel 3 foot 6¾ inch diameter by 10 feet 1 inch in length, height of centre line from rails 5 feet 10½ inches. The extreme length of the engine was 24 feet 6 inches, the water capacity of the tank was 625 gallons, and the weight in working order was 25 tons 12 cwt, 7 tons 6 cwt being on the leading wheels, 9 tons 9 cwt on the driving and 8 tons 17 cwt on the trailing. From our illustration it will be noticed that these engines had inside bearings for the driving, but outside for the leading and trailing wheels. There was a water tank under the footplate, and generally they bore a striking resemblance to a

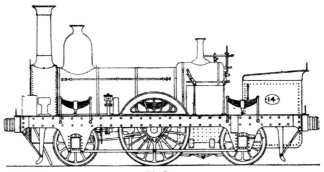

Fig. 21.

'Sharp'. No. 14 was re-numbered 278 in June, 1874, and broken up in March, 1878.

The next locomotive built at Brighton was a single-wheel express engine, completed in December, 1852, and numbered 48. This engine had inside bearings for the driving and outside for the leading and trailing wheels, and was generally after the design of the 'Jenny Lind', with the exception of the steam dome, which was placed close against the chimney. We are unable to illustrate this engine, but some of its chief dimensions were: Cylinders 15 inch diameter by 22 inch stroke, diameter of leading and trailing wheels 4 feet, diameter of driving wheels 6 feet 1 inch, leading to driving and driving to trailing centres each 7 feet, making a total wheel-base of 14 feet; external diameter of boiler 3 feet 8¾ inches, length 10 feet 6 inches, length of tubes 10 feet 11¼ inches, length of outside firebox 4 feet 2¼ inches, weight of engine 26 tons 2 cwt, the leading wheels carrying 10 tons, the driving 11 tons 6 cwt, and the trailing 4 tons 16 cwt. The boiler was constructed in three rings, having a lock-up safety valve on the steam dome and spring balance safety valves over the firebox. No. 48 was re-numbered 274 in October, 1876, and in December, 1878, 412, and was scrapped in June, 1881.

These three engines were the only ones built in 1852, and a similar number were constructed in 1853, one of which was a small single express, No. 24. Very little appears to be known of the general appearance of this engine, but a few of its dimensions were: Cylinders 14 inch diameter by 22 inch stroke, diameter of driving wheels 5 feet 6 inches, boiler 3 foot 6 inch diameter by 9 feet 6 inches long, containing 133 tubes 2 inch diameter by 10 feet in length. This engine only weighed 20 tons 10 cwt, having 7 tons 16 cwt on the leading wheels, 9 tons 12 cwt on the driving, and 3 tons 2 cwt on the trailing – a distribution of weight which allowed very little for the trailing end.

The next engines coming under our notice are four single expresses, illustrated by *Fig. 22*, numbered 10, 23, 38 and 41. The last-mentioned was built in July, 1853, and No. 10 in September, 1853, but Nos 23 and 38 were not completed until July, 1854. In detail and construction they were practically identical with No. 48, but differed slightly in dimensions. The cylinders were 14 inch diameter by 22 inch stroke, diameter of driving wheels 5 feet 6 inches, boiler 3 foot 6 inch diameter by 10 feet long, containing 143 tubes 2 inch diameter by 10 feet 7 inches in length. These engines weighed 26 tons 10 cwt, of which the leading wheels carried 10 tons 2 cwt, the driving 10 tons and the trailing 6 tons 8 cwt. No. 10 was re-numbered 290 and broken up in March, 1878;

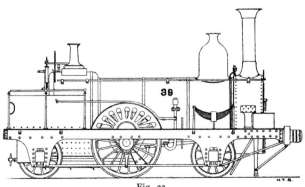

Fig. 22.

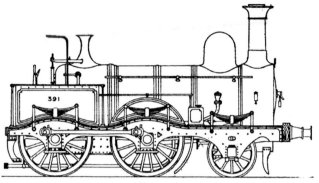

Fig. 23.

No. 23 was re-numbered 266 and again No. 484.

We now come to two six-wheels coupled engines, numbered 44 and 46, which were the first goods engines built at Brighton Works, No. 44 being turned out in January, 1854, and No. 46 in March that year. They were both the same in construction and detail, except that while No. 46 had only one steam dome, No. 44 had two, one in the usual position on the boiler and the other over the firebox, and was a very unsightly engine, the domes being very high and reaching nearly to the level of the chimney top. These two engines had iron frames and inside bearings to all the wheels, and their chief dimensions were as follows: Cylinders 16 inch diameter by 24 inch stroke, diameter of blast pipe orifice 4 inches, diameter of wheels 4 feet 9 inches, leading to driving centres 7 feet 2 inches, driving to trailing 7 feet 10 inches, total wheel-base 15 feet; boiler, made in three rings, butt jointed, with the steam dome on the first ring, external diameter 3 feet 9 inches, length 11 feet 6 inches, number of tubes 156, diameter of tubes 2 inches, length 12 feet 1 inch, length of outside firebox 4 feet 9 inches, width 4 feet, height of centre line of barrel from rail level 6 feet 3 inches, height of chimney 13 feet 4 inches, weight on leading wheels 10 tons 12 cwt, on driving 12 tons 4 cwt, on trailing 6 tons 12 cwt, total weight 29 tons 8 cwt. A spring balance safety valve was fixed on each of the domes of No. 44, the regulator being in the dome on the boiler, while No. 46 had both safety valves over the firebox. No. 44 was rebuilt in October, 1864, and No. 46 in March of the same year. In July, 1872, they were re-numbered 271 and 272, and at a later date 303 and 304. In December, 1877, they were again re-numbered 401 and 403. No. 403 was scrapped in March, 1878.

Eight engines, built for mixed traffic work, were purchased by the company in four different lots, two at a time. These engines had the same size of cylinders and wheels, and were of the same external appearance, the difference in them, if any, being slight *(Fig. 23)*. They were purchased from Sharp, Stewart & Co. (formerly Sharp, Brothers & Co.), of the Atlas Works, Manchester, and were numbered 6, 13, 16, 20, 40, 42, 100 and 116. The first lot, Nos 16 and 20, were bought in September, 1854, and the next, Nos 13 and 40, in November, 1854, the maker's Nos of the last-named being 882 and 883. In July, 1855, Nos 6 and 116 were purchased, their maker's Nos being 925 and 926, and the last two, Nos 42 and 100, were bought in April, 1856. The following are the principal dimensions of Nos 16 and 20, which may be taken as representing the whole class: Cylinders 16 inch diameter by 20 inch stroke, distance between centres 2 feet 5 inches, diameter of leading wheels 3 feet 6 inches, diameter of driving and trailing wheels 5 feet 6 inches, leading to driving centres 6 feet 6 inches, driving to trailing 7 feet, total wheel-base 13 feet 6 inches; boiler barrel, made in two rings, with the steam dome on the first, external diameter 4 feet, length 9 feet 9 inches, number of tubes 170, diameter of tubes 2 inches, length 10 feet 1½ inches, height of centre line of boiler from rails 6 feet 2½ inches, length of outside firebox 4 feet 1¼ inches, width 4 feet 3¾ inches, depth below centre line of boiler 4 feet 5½ inches, heating surface of tubes 905.5 sq. ft, of firebox 69.5 sq. ft, total 975 sq. ft, working pressure 120 lbs per sq. in., weight on leading wheels 8 tons 14 cwt, on driving 9 tons 6 cwt, on trailing 6 tons 16 cwt, total weight 24 tons 16 cwt. These engines had sandwich framing outside, and iron frames inside, a lock-up safety valve on the dome, and spring balances over the firebox.

Engine No. 40 was rebuilt in June, 1871, by Stroudley, who put in a new boiler of his own design, and also added the standard cab over the footplate, and hung the driving and trailing springs under the axle boxes, instead of above as they were originally, whilst new driving splashers and sandbox combined were fitted up, the sand gear being worked by a lever from the cab. This engine, after it came out rebuilt, was named *Epsom*, and ran for some time between that town and London, and was finally sold in December, 1891, after having been working for a period of thirty-seven years. Stroudley did not rebuild any other engines of this class, all having been scrapped a

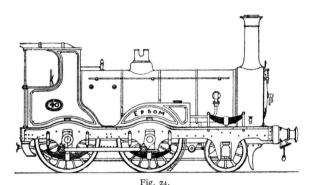

Fig. 24.

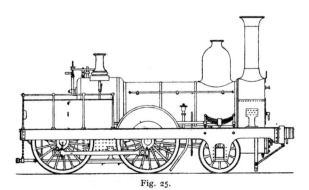

Fig. 25.

long time back. An illustration of No. 40, as rebuilt by Stroudley, is shown in *Fig.* 24. The new boiler was made in two rings, its external diameter being 4 feet, 2 inches, and its length 8 feet 8 inches. It contained 236 tubes 1½ inch diameter, and 9 feet 2 inches long, the firebox being 5 feet 2 3/16 inches in length by 4 feet in width, and its depth from the centre line of the boiler 4 feet 4¼ inches. The heating surface of the tubes was 849 sq. ft, and firebox 89 sq. ft, making a total of 938 sq. ft. Like all of Stroudley's earlier engines, when first rebuilt it had no steam dome, but Adams' patent safety valves over the manhole. A few years later this arrangement was done away with, and a steam dome substituted with Salter's spring balance safety valves. After these alterations the weight was increased by nearly 5 tons, there being on the leading wheels 9 tons 4 cwt, on the driving 12 tons, and on the trailing 8 tons 8 cwt, making a total of 29 tons 12 cwt.

No. 6 was re-numbered 289, and in June, 1879, 390, and scrapped in March, 1882. No. 13 was scrapped in April, 1886. No. 20, in June, 1875, was re-numbered 264. No. 40 was re-numbered 260 in January, 1878, and in October, 1881, 464. No. 42 was re-numbered in June, 1877, 355, and scrapped in March, 1882. No. 100 was re-numbered 257, and in October, 1881, 505, and broken up in September, 1882. No. 116 was re-numbered 356 in June, 1877, and scrapped in June, 1881.

The first coupled passenger tender engines built by Craven at Brighton were two for working the traffic between London and Croydon, which were always spoken of as the 'Croydon engines'. They were numbered 1 and 2, and were built in November, 1854. *Fig.* 25 gives an illustration of this class, from which it will be seen that they had inside bearings to the coupled wheels, and outside ones to the leading pair of wheels. No. 2 was the first locomotive on this railway to have a steam blower, the pipe, instead of being placed inside the boiler as is now usual, being fixed on the outside and entering the smokebox close to the steam pipe. To prevent condensation in this pipe, it was lagged round with wood, which gave the engine a very ugly appearance. The principal dimensions of Nos 1 and 2 were as follows: Cylinders 15 inch diameter by 22 inch stroke, diameter of leading wheels 3 feet 6 inches, diameter of coupled wheels 5 feet 6 inches, leading to driving centres 6 feet 3½ inches, driving to trailing 7 feet 6 inches. The boilers were made in three rings, having the steam dome with a lock-up safety valve on the first, and spring balances over the firebox. The external diameter of the boiler was 3 feet 8 inches by 10 feet in length, the centre line being 5 feet 10½ inches from the rails. The length of the outside firebox was 3 feet 9⅞ inches, and its width 4 feet. There were 147 2 inch tubes 10 feet 5½ inches long. The heating surface of the tubes was 742 sq. ft, and of the firebox 70 sq. ft, making a total of 812. These engines weighed 25 tons 7 cwt, the leading wheels carrying 9 tons 5 cwt, the driving 10 tons 10 cwt,, and the trailing wheels 5 tons 12 cwt. No. 2 was re-numbered 286, and in June, 1879, 389, and scrapped in May, 1880.

In November, 1854, the LBCSCR purchased from the Manchester, Sheffield & Lincolnshire Railway two very heavy and powerful goods engines which had been built for the latter by Sharp, Stewart & Co., but finding them too powerful for their requirements, they sold them to the Brighton Company, who were then in want of some goods engines for working heavy ballast trains. These two locomotives were named

Orestes and *Europa*, and numbered 117 and 121 respectively, these numbers being retained when they came to Brighton. They were painted a chocolate brown colour and had a brass number plate on the front of the chimney, the name plate on the side of the boiler being also of brass. For a considerable time after they came they used to run trains of chalk, consisting of fifty loaded wagons out, and of seventy wagons, some loaded and some empty, back. *Fig. 26* shows one of these engines, of which the chief particulars are: Cylinders 18 inch diameter by 24 inch stroke, distance apart of centres 2 feet 3 inches, inclination 1 in 10, the slide valves being underneath the cylinders; diameter of blast pipe orifice 4½ inches, diameter of wheels 5 feet 0½ inch, wheelbase 14 feet, being 7 feet between each pair of centres, and the driving and trailing springs being connected by a compensating lever. The boilers were made in three rings, length 11 feet, external diameter 4 feet 4 inches, height of centre line above rail 6 feet 6 inches, number of tubes 229, length 11 feet 7 inches, diameter 2 inches, length of outside firebox 6 feet 8 inches, width 4 feet 2¾ inches, depth below centre line of boiler 4 feet 9 inches, heating surface of tubes 1,426.9 sq. ft, and of firebox 145.6 sq. ft, giving a total surface of 1,572.5 sq. feet. They had spring balance safety valves over the firebox, and the regulator was not in the dome, but in the boiler at the smokebox end. The weight of these engines was 33 tons 10 cwt, the leading wheels having 9 tons 4 cwt, the driving 13 tons 10 cwt, and the trailing 10 tons 16 cwt. It may also be noted that the coupling and connecting rods were round in section. From these dimensions the size and power of the engines will be seen at once, and we should think that they were about the most powerful engines for ordinary work on any British railway at that time. The centre of the trailing axle was placed under the centre of the firebox, the large dimensions of which will be noticed.

In February, 1870, No. 117, and in March, No. 121 were rebuilt with new boilers constructed for burning coal, coke having been the fuel previously used. These boilers were not quite so large as the original ones, but the firebox was larger and made double, some of their principal dimensions being as follows: Barrel, made in three rings, length 10 feet, diameter 4 feet 2 inches, length of firebox 7 feet, depth at front below centre of boiler 4 feet 6 inches, inclined at the back to a depth of 1 foot 10½ inches, number of tubes 157, diameter of tubes 1¾ inches. A spring balance safety valve was put on the dome and a lock-up valve over the firebox, and the diameter of

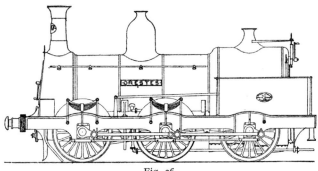

Fig. 26.

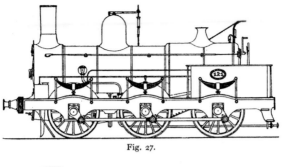

Fig. 27.

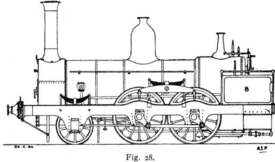

Fig. 28.

the blast pipe was increased from 4½ inches to 5¼ inches, their consumption of fuel having been very heavy. *Fig.* 27 illustrates them after their alteration, from which it will be seen that their names were removed and that they were considerably altered in external appearance. For a long time after they came out with the new boilers they ran without any lagging on. When, some years after, engines were built with their numbers, they were re-numbered 362 and 370, and the former engine was again re-numbered 398. In September, 1885, No. 398 was scrapped, and in April, 1886, No. 370 was put out of service. It should be remarked that these new boilers were put in under Craven's orders, but that gentleman having left the service of the company before they were finished, they were turned out under Stroudley's superintendence. These were spoken of as the finest goods engines the LBSCR then had. The company thus added twelve engines to its stock during the year 1854, having built six and purchased six.

In 1855 Craven designed and constructed at the Brighton Works two classes of four-wheels coupled 'long boiler' tender engines, one class for working mixed traffic, and the other for passenger service. There were, however, only two locomotives to each class, the first constructed being the mixed traffic engines, which were numbered 3 and 5 and turned out in February and March, 1855. The passenger engines were numbered 8 and 9 and built in July, 1855. In December, 1856, two more engines of this type were constructed, but they were of larger dimensions and were numbered 120 and 124.

We first give the principal particulars and dimensions of engines Nos 3 and 5: Cylinders 16 inch diameter by 22 inch stroke, centres of cylinders 2 feet 7 inches; diameter of leading wheels 3 feet 6 inches, diameter of driving and trailing wheels 5 feet, leading to driving centres 6 feet 2½ inches, driving to trailing 5 feet 3 inches; boiler, made in four rings, external diameter 3 feet 9 inches, length 13 feet, height of centre line from rail level 5 feet 8½ inches; outside firebox, length 4 feet 2 inches, breadth 4 feet, depth below centre of boiler 4 feet 7 inches. These engines had outside bearings with sandwich frames, and plate frames inside. Their weight was 28 tons 14 cwt, distributed as follows: On leading wheels 7 tons 13 cwt, on driving 10 tons 14 cwt, and on trailing 10 tons 7 cwt. Engine No. 3 was re-numbered 287, and scrapped in April, 1879.

Fig. 28 shows engines Nos 8 and 9, their chief dimensions being: Cylinders 15 inch diameter by 20 inch stroke, centres of cylinders 2 feet 7 inches, diameter of blast pipe orifice 4¾ inches, diameter of leading wheels 3 feet 6 inches, diameter of driving and trailing wheels 5 feet 6 inches, leading to driving centres 6 feet 6 inches, driving to trailing 5 feet 11 inches; boiler, constructed in four rings with the steam dome on the third, outside diameter 3 feet 8⅛ inches, length 13 feet 6 inches, height of centre line from rails 5 feet 9 inches, length of tubes 14 feet; outside firebox, length 4 feet 1½ inches, breadth 4 feet These engines had a vertical revolving disc valve regulator in the steam dome, which had the usual lock-up safety valve, there being also a spring balance over the firebox. They had outside sandwich frames, and iron plate frames inside between the cylinders and firebox, the driving and trailing springs being connected with a compensating lever. The throw of the side rods was 12 inches, and the weight of the engines as follows: On leading wheels 7 tons 12 cwt, on driving 10 tons 10 cwt, on trailing 10 tons 10 cwt; total 28 tons 12 cwt.

Engines Nos 120 and 124, as we have already mentioned, were of similar build to Nos 8 and 9, but were larger, and for comparison their chief dimensions were: Diameter of cylinders 16 inches, stroke 22 inches, centres of cylinders 2 feet 7 inches, diameter of leading wheels 4 feet, diameter of driving and trailing wheels 5 feet 8 inches, leading to driving centres 6 feet 6 inches, driving to trailing 5 feet 11 inches; boiler, made in four rings, external diameter 3 feet 9 inches, length 13 feet 6 inches, height of centre line above rail level 5 feet 8½ inches; outside firebox, length 4 feet 1 inch, breadth 4 feet, depth below centre of boiler 4 feet 4 inches. These engines also had outside sandwich frames and iron plate frames inside, and their weight was: On leading wheels 7 tons 14 cwt, on driving 11 tons 4 cwt, on trailing 11 tons 7 cwt; total 30 tons 5 cwt.

The next class of engine coming under our notice is one consisting of two curious little front-coupled saddle tank engines, built for working branch line trains. They were

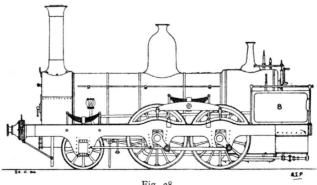

Fig. 28.

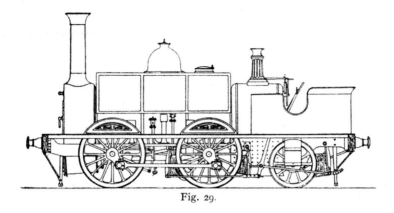

Fig. 29.

numbered 11 and 22, and were built in July and August, 1855, respectively, *(Fig. 29)*. The most peculiar feature of these engines was the framing. From the leading buffer beam to a point between the cylinders and the leading axle the frames were made of iron plate, from there to the front plate of the firebox the framing was built up of iron bars in a similar manner to the Bury engines. From the front to the back of the firebox there was no framing at all, but from there to the trailing buffer beam there was a thin iron plate frame. Another curious point about these two engines was that the cylinders were inclined upwards to the driving axle and that the piston rods were underneath the leading axle, the steam chests being placed at an angle of about 45 degrees to the vertical. The slide bars were round, 2½ inches in diameter, and 4½ inches between top and bottom bars; the connecting rods 5 feet 6½ inches long and forked for about half their length in order to clear the slide bars. Their principal dimensions were: Cylinders 15 inch diameter by 20 inch stroke, distance apart of centres 1 foot 11 inches, diameter of blast pipe orifice 4⅞ inches, diameter of leading and driving wheels 5 feet, diameter of trailing wheels 3 feet 6 inches, leading to driving centres 7 feet 5 inches, driving to trailing 6 feet 6½ inches, total wheelbase 13 feet 11½ inches; boiler, made in three rings, external diameter 3 feet 8 inches, length 10 feet 6 inches; outside firebox, length 4 feet, breadth 4 feet 3 inches, depth below centre line of boiler 4 feet 0¼ inch, 125 brass tubes 2 inch diameter by 11 feet long; heating surface, tubes 625 sq. ft, firebox 68 sq. ft, total 693 sq. ft. There was a lock-up safety valve on the dome, and two spring balance valves over the firebox. These engines weighed 28 tons 17 cwt, distributed as follows: On leading wheels 11 tons 16 cwt, on driving wheels 10 tons 6 cwt, and on trailing wheels 6 tons 15 cwt. The saddle tank was square topped and contained 600 gallons, of water; the trailing springs were of the volute pattern, resting on the top of the axle boxes, the other wheels being fitted with ordinary springs.

We now come to a class of six-wheels coupled goods locomotive, of which five engines were built at the Brighton Works, though not all at one time. They were numbered 110, 118, 119, 134 and 143; Nos 118 and 119 being built in November, 1855, No. 134 in January, 1859, No. 143 in January, 1861, and No. 110 in November, 1862. *Fig. 30* illustrates No. 118, their principal dimensions being: Cylinders: diameter 16 inches, stroke 24 inches, distance apart of centres 2 feet 6 inches; diameter of wheels 5 feet,

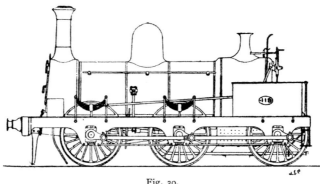

Fig. 30.

wheel-base 15 feet 6 inches, being 7 feet 4 inches between leading and driving centres, and 8 feet 2 inches between driving and trailing; boiler, made in three rings, external diameter 4 feet, length 11 feet 6 inches, height of centre line from rail level 6 feet 6 inches, number of tubes 158, diameter of tubes 2 inches: outside firebox, length 4 feet 8½ inches, breadth 4 feet 3 inches, depth below centre line of boiler 5 feet 4¼ inches. A plain steam dome was placed on the middle of the boiler, containing a steam pipe leading to the smoke-box, at which end of the boiler the regulator was situated, and two spring balance safety valves were fixed over the firebox. The latter was a double one, having a 4 inch water space between the inner boxes, and a heating surface of 121 sq. ft, that of the tubes being 979 sq. ft, giving a total of 1,100 sq. ft. These engines had sandwich frames outside and iron plate frames inside from the leading buffer beam to the front of the firebox, and weighed 30 tons 18 cwt, having 11 tons on the leading wheels, 12 tons 18 cwt on the driving and 7 tons on the trailing wheels.

In February, 1877, No. 110 was re-numbered 353, being finally scrapped in December, 1878, while No. 134 was re-numbered 384 in December, 1878, and was purchased in July, 1885, by the West Lancashire Railway Company.

In 1856, Craven designed and constructed four single wheel express engines for working fast trains between London and Brighton, and Portsmouth. The two first of these were designed practically on the lines of Sharp's large single engines, and in general appearance were almost identical with them, as will be seen from *Fig. 31*. They were numbered 125 and 126, and were built in July, 1856. The other two were after the design of the Wilson locomotives, were numbered 122 and 123, and were turned out in September, 1856. Nos 125 and 126 were named by Stroudley *Ventnor* and *Shanklin* respectively, and in July, 1878, re-numbered 372 and 373, whilst No. 123, *(Fig. 32)*, was named *Drayton*. Nos 372 and 373 were scrapped in June, 1881.

Although these two classes were like the Sharp and Wilson engines in appearance, they were of different dimensions, those of engines Nos 125 and 126 being: Cylinders: diameter 16 inches, stroke 22 inches; diameter of leading and trailing wheels 4 feet, diameter of driving wheels 6 feet 6 inches, leading to driving centres 7 feet 2½ inches, driving to trailing 7 feet 4 inches, total wheel-base 14 feet 6½ inches; boiler, made in three rings, external diameter, 3 feet 11 inches, length 11 feet 5 inches, and containing 162 brass tubes of 2 inch diameter; length of outside firebox 4 feet 1 inch. The regulator was of the revolving disc type, and was situated in the steam dome, which was on the first ring of the barrel. The heating surface was: Tubes 976 sq. ft, firebox 73 sq. ft, total 1,049 sq. ft; and the weight was 26 tons 16 cwt, distributed thus; on leading wheels 9 tons 4 cwt, on driving wheels 12 tons 14 cwt, and on trailing wheels 4 tons 18 cwt.

The following particulars relate to engines Nos 122 and 123: Cylinders: diameter 15 inches, stroke 20 inches; diameter of leading and trailing wheels 4 feet, diameter of driving wheels 6 feet, wheel-base 14 feet, equally divided; boiler, made in three rings, with the steam dome on the middle one, external diameter 3 feet 10 inches, length 10 feet, height of centre line from rails 6 feet 4 inches; the firebox was double, having a length of 4 feet 9 inches outside, and the regulator was in the boiler, steam being conveyed to it by a vertical pipe in the steam dome, surrounded by a pan to prevent the water from getting to the regulator. These engines weighed 26 tons 6 cwt, having 9 tons 8 cwt on the leading wheels, 11 tons 18 cwt on the driving wheels and 5 tons on the trailing wheels.

Engines Nos 122 and 126 were in the lamentable Clayton tunnel accident of 1861. On the morning of Sunday, 25 August, an excursion train left Portsmouth at 6 a.m. for London, via Brighton, being booked to leave the latter station at 8.05. An excursion from Brighton to London was due to leave at 8.15, and the ordinary 'Parliamentary' train at 8.30. About 4 miles from Brighton is the Clayton tunnel, 1 mile 22 chains in length, at both ends of which were signal boxes, and it was then, as now, the rule that no two trains should be in the tunnel at the same time on the same line of rails. The Portsmouth train, drawn by engine No. 48, passed through the tunnel safely, but as the signalman at the south end had not heard from the man at the north end when the Brighton excursion arrived, drawn by engine No. 126, this should have been stopped

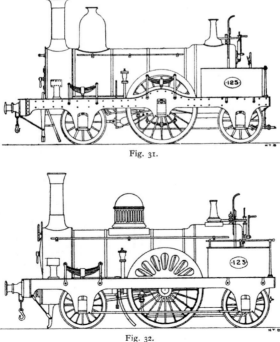

Fig. 31.

Fig. 32.

by the fixed signal at the south end of the tunnel. This, however, failed to act, and the signalman endeavoured to stop the driver by showing him a red flag, but he was unable to do this at once, owing to the speed of the train, and the signalman, doubtful whether his red flag had been seen by the driver, telegraphed to the north end man to know if the train had passed out. This signalman, supposing the inquiry to refer to the Portsmouth train, replied that it had, whereupon the south end man, taking it to mean that the Brighton excursion had passed through, displayed a white flag to the ordinary train, which was drawn by No. 122, entered the tunnel at full speed. In the meantime the driver of the Brighton excursion, having seen the signalman's red flag, had brought his train to a stand in the tunnel, and was slowly backing it towards the southern outlet, when it was dashed into by the ordinary train, the engine of the latter mounting the two rear vehicles of the excursion, which it completely demolished, and then fell forward and remained in a vertical position, resting on its leading end. The carriages were unfortunately full of passengers, some of whom were crushed and mangled and some burned, twenty-three lives being lost in all, and 176 persons being injured.

In 1858, six four-wheels coupled tank engines were constructed at Brighton from Craven's designs for working that portion of the system known as the West End of London and Crystal Palace Railway. These locomotives were of a rather peculiar design with regard to the arrangement of the cylinders and frames. As will be seen from *Fig. 33*, they were six-wheeled, having outside frames, with the driving and trailing wheels coupled. They had four framings in all, the cylinders being fixed between the inside and outside ones. Underneath the barrel of the boiler and between the inside framings was placed a well tank, and another was under the coal bunker behind the trailing wheels. The boilers of these engines had two steam domes, one in the usual position on the barrel and the other over the firebox, a spring balance safety valve being fixed on each. The feed pumps were fastened to the outside frames, the plungers being actuated by an eccentric and rod attached to the coupling rod cranks. These engines had two slide bars, round in section, one at the top and one at the bottom, and their principal dimensions were: Cylinders, 15 inch diameter by 22 inch stroke, diameter of leading wheels 3 feet 4 inches, diameter of driving and trailing wheels 5 feet 6 inches, wheelbase 13 feet 6 inches, being 6 feet 9 inches between each pair of centres. The diameter of the boiler barrel was 3 feet 10 inches, and its length 10 feet

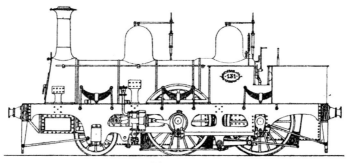

Fig. 33.

5 inches, the height of the centre line above the rail level being 5 feet 7¼ inches. There were 108 tubes of 2¼ inch diameter and 10 feet 9½ inches long, and the length of the outside firebox was 4 feet 1½ inches. The total heating surface was 764 sq. ft, that of the tubes being 682 sq. ft and of the firebox 82 sq. ft. The well tank under the boiler was capable of holding 178 gallons, and that under the footplate 364 gallons, giving a total of 542 gallons. The weight of these engines was 35 tons 19 cwt, having the weight distributed as follows: 10 tons 6 cwt on the leading wheels, 13 tons 5 cwt on the driving and 12 tons 8 cwt on the trailing.

The engines were numbered 12, 15, 105, 107, 128 and 129 and were built, re-numbered and broken up as follows:

No.	Date	Re-No. and Date	Re-No. and Date	Re-No. and Date	Scrapped
12	May 1858	131	378, Sept 1878		1889
15	April 1858	106	292	367, Oct 1877	July 1882
105	Feb 1858	291			
107	April 1859	281	393, Oct 1879		June 1882
128	June 1858	185			
129	May 1858	376, Sept 1878			May 1880

Soon after No. 12 was built it was altered to 131, for at this time the previous No. 12, which had been condemned and was waiting to be broken up, was considered, upon a further examination, to be worth slight repairs, and would therefore remain in service some time longer. It will be noticed from the illustration that these engines appeared to have a slightly raised firebox; this, however, was not the case, the sheet iron cleading being raised for some reason we have been unable to discover. Some few years later Craven reconstructed these engines as ordinary side tanks, of which *Fig. 34* is an illustration. A pair of cylinders 16 inches by 20 inches were put in and placed between the inside frames in the ordinary manner, and the 3 foot 4 inch leading wheels were replaced by a pair 4 feet in diameter. The driving and trailing springs were hung under the axle boxes and the pumps placed inside the frames. New outside frames were provided, and the wheel-base lengthened 18 inches, the distance between the centres being 7 feet 9 inches and 7 feet 3 inches respectively. Four square slide bars took the place of the two round ones, and the steam dome over the firebox was

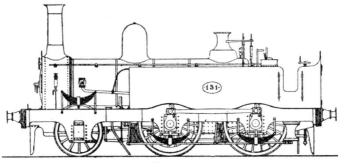

Fig. 34.

dispensed with, a manhole cover being substituted, on which were mounted the safety valves. The capacity of the side tanks was 560 gallons and of that under the footplate 200 gallons. These engines were slightly heavier after their alteration, weighing about 36 tons 10 cwt each.

In September, 1858, a small four-coupled tender engine was built at Brighton for working passenger trains between London and Croydon, and was numbered 101. The cylinders were 16 inch diameter, with a stroke of 20 inches, the centres being 2 feet 5 inches apart. The diameter of the leading wheels was 3 feet 6 inches, and of the driving and trailing wheels 5 feet 6 inches. The wheel-base was 13 feet 9 inches, being 6 feet 6 inches from leading to driving centres and 7 feet 3 inches from driving to trailing. The barrel of the boiler was made in three rings, its external diameter being 3 feet 9¼ inches and its length 10 feet, the height of the centre line from the rails being 6 feet 1 inch. The outside firebox was 4 feet 9¾ inches in length and 4 feet 2½ inches in width, the inside firebox having a depth of 4 feet 8¾ inches and a partition or midfeather running from the front to the back. This engine had outside sandwich frames, and iron plate inside frames extending from the front buffer beam to the front plate of the firebox, to which they were connected by cast-iron brackets; all the wheels had outside bearings. The weight was 27 tons 10 cwt, the leading wheels carrying 9 tons, the driving 11 tons 2 cwt and the trailing 7 tons 8 cwt. No. 101 was subsequently re-numbered 258, and in October, 1881, 506, this number being the highest on the line if we except Bury's old No. 99, re-numbered to 9⁹9, as previously mentioned. In Stroudley's administration this engine was named *Rouen*, the original open splashers were replaced with closed-in ones, and the name painted on. The driving and trailing springs were then underhung, instead of being above the footplate. No. 506 was scrapped in January, 1885.

On 18 October, 1858, the line from Lewes to Uckfield was opened to the public, and for working the service on this branch Craven built a small four-coupled tank engine, which was numbered 130, and is illustrated by *Fig. 35*. The chief particulars of this engine were: Cylinders 13 inch diameter by 18 inch stroke, centres 2 feet 4 inches apart; diameter of leading wheels 3 feet 6 inches, and of driving and trailing wheels 5 feet 1½ inches, the springs of the coupled wheels being connected by a compensating lever; wheel-base 13 feet 10½ inches, the distance between leading and driving centres being 5 feet 11½ inches, and between driving and trailing 7 feet 11 inches. The barrel of the boiler was made in three rings, and had an external diameter of 3 feet 4¾ inches, and a length of 8 feet 11½ inches, its centre line being only 5 feet 4¾ inches above the rail level. The length of the outside firebox was 4 feet 9 inches, and the width 3 feet 10 inches. The inside firebox had a depth of 3 feet 10½ inches, and was of iron, iron bolts also being used for staying the sides, instead of the usual copper firebox and stays. The boiler contained 113 brass tubes, having a heating surface of 550 sq. ft, that of the firebox being 70 sq. ft, making a total of 620 sq. ft. The regulator was in the barrel of the boiler at the smokebox end, the steam pipe leading up into the dome, which was on the second ring of the barrel, and two spring balance safety valves were fixed over the firebox. As will be seen from the illustration, the side tanks were extremely short, and together only carried 193 gallons of water; a tank was also placed under the footplate behind the trailing wheels to hold 237 gallons, thus giving a total water capacity of

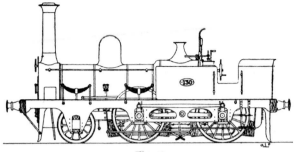

Fig. 35.

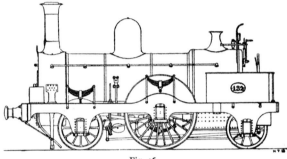

Fig. 36.

430 gallons. This engine had outside sandwich frames, and inside iron plate frames between the cylinders and front plate of the firebox, and weighed 28 tons 16 cwt, distributed as follows: On leading wheels 8 tons 16 cwt, on driving 10 tons 4 cwt, and on trailing 9 tons 16 cwt. No. 130 was scrapped in August, 1877.

In January, 1859, two single-wheel express engines were constructed at the company's works at Brighton from designs prepared by Craven. These two locomotives were numbered 132 and 133, but they differed both in design and dimensions. No. 132 was a large single, with outside bearings to all the wheels, whilst No. 133 was a small single with inside bearings to the driving wheels. The former engine is illustrated by *Fig. 36*, and the particulars of it are as follows: Diameter of cylinders 16 inches, stroke 22 inches, diameter of leading wheels 4 feet, of driving wheels 6 feet 6 inches, and of trailing wheels 3 feet 8 inches, leading to driving centres 7 feet 4½ inches, driving to trailing 7 feet 11 inches, total wheelbase 15 feet 3½ inches, external diameter of boiler 3 feet 11¾ inches, length 11 feet 6 inches, height of centre line from rail level 6 feet 6 inches, 138 brass tubes of 2 inch diameter, length of outside firebox 4 feet 8½ inches, breadth 4 feet 3 inches, depth of inside box 5 feet 4¼ inches. This engine had a double firebox, and the inside box was connected to the outer shell in the same manner as in No. 130. The safety valves were placed over the firebox, and the regulator was in the boiler, as was the case with the majority of these engines. No. 132 had outside sandwich frames and inside iron plate frames between the cylinders and the firebox, the inside frames being placed between the inside webs of the cranks, the driving axle having two bearings between the cranks and the eccentric sheaves. This engine had the large amount of 1,100 sq. ft of heating surface, the tubes giving 979 sq. ft and firebox

121 sq. ft. Its weight was 29 tons 6 cwt, the leading wheels supporting 9 tons 12 cwt, the driving 13 tons 10 cwt, and the trailing 6 tons 4 cwt.

Engine No. 133 is illustrated by *Fig. 37*, its dimensions being: Diameter of cylinders 15 inches, stroke 22 inches, diameter of leading wheels 4 feet, of driving wheels 5 feet 6 inches, and of trailing wheels 3 feet 6 inches, leading to driving and driving to trailing centres each 7 feet 1 inch, total wheelbase 14 feet 2 inches; boiler, made in three rings, external diameter 3 feet 7 inches, length 10 feet 5 inches, height of centre line above rail level 6 feet 1½ inches, 122 brass tubes of 2 inch diameter. The steam dome was placed on the middle ring of the boiler, one safety valve being fixed thereon, and another over the firebox, and the regulator was at the smokebox end of the barrel. The firebox casing was 4 feet 3 inches in length, and was bolted to the inside box similarly to engine No. 132. The reversing lever was placed outside the driving splasher, instead of going behind it and under the firebox lagging. This engine had outside sandwich frames, and inside frames running between the back of the cylinders and front of the firebox, and its weight was 25 tons, the leading wheels having 9 tons 1 cwt, the driving 11 tons 1 cwt, and the trailing 4 tons 18 cwt. As will be seen from the illustration, No. 133 was similar in appearance to the 'Jenny Linds', and after it had been modernized by having sheet-iron lagging and other details introduced by Craven, it was in fact taken for such by many; it was latterly named *Penge*, and re-numbered in November, 1878, 405, being scrapped in April, 1879.

In April, 1859, three six-wheels coupled goods tender engines were turned out of the shops at Brighton, numbered 102, 107 and 135 *(Fig. 38)*. They were pretty little engines, with the exception of their wheels, which were of cast-iron, with rims of T-section. As a matter of fact, the wheels and motion of these locomotives had been previously used in the goods engines built by Longridge in 1847, previously described. Their principal dimensions: Diameter of cylinders 16 inches, stroke 24 inches, diameter of wheels 4 feet 9 inches, leading to driving centres 6 feet 7 inches, driving to trailing 7 feet 5 inches, total wheel-base 14 feet; external diameter of boiler 3 feet 9¾ inches, length 10 feet 6 inches, height of centre line from rail level 6 feet 3¼ inches, number of tubes 152, diameter of tubes 2 inches, length of outside fire-box 4 feet 6½ inches, breadth 3 feet 11 inches, depth of inside box 4 feet 1½ inches. The steam dome was placed on the

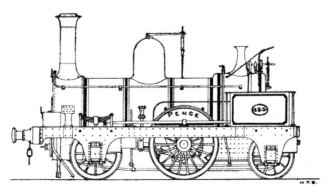

Fig. 37.

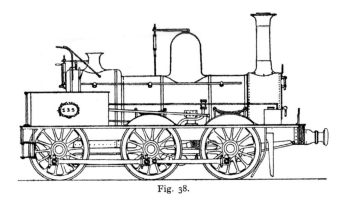

Fig. 38.

second ring of the boiler, and the regulator was in the latter at the smoke-box end, two spring balance safety valves being placed over the firebox. They had only one pair of frames, which were of iron, and inside or between the wheels, and they weighed on the average 28 tons 18 cwt, the leading wheels carrying 11 tons 3 cwt, the driving 11 tons 5 cwt, and the trailing 6 tons 10 cwt. No. 102 was first re-numbered 259, then 311 in November, 1875, and in May, 1883, 500, and was scrapped in January, 1884. No. 107 was re-numbered 281, and in October 1879, 393, and was scrapped in June, 1882. No. 135 was scrapped in June, 1877. The illustration shows No. 135 with new wheels of the ordinary type.

In June, 1859, an experimental engine was built at the company's works for the West End of London line, and was numbered 136. It was an outside cylinder tank, and was, we believe, the first tank engine combining in its design a leading bogie, four-coupled wheels and outside cylinders. It is illustrated by *Fig. 39*, from which it will be noticed that a striking peculiarity was the great inclination of the cylinders, which were 15 inches in diameter with a stroke of 22 inches, the centres being 6 feet 4½ inches apart. The diameter of the bogie wheels was 3 feet, and of the driving and trailing 5 feet. The total wheel-base was 16 feet 9½ inches, the centres of the bogie wheels being 4 feet apart, from the centre of the bogie to the driving centre was 8 feet 0½ inches, and from driving to trailing centres was 6 feet 9 inches. The bogie had bar framing with inside

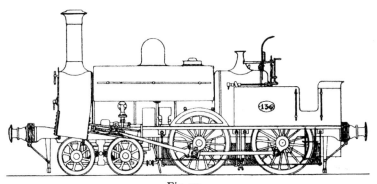

Fig. 39.

bearings. The external diameter of the boiler was 3 feet 9 inches, and its length 10 feet; it contained 132 brass tubes of 2 inch diameter, and 10 feet 5¼ inches in length. The length of the outside firebox was 4 feet 3 inches, its width 3 feet 10 inches, and the depth of the inside box 4 feet 11⅛ inches. The tubes afforded 733 sq. ft of heating surface, and the firebox 71.8 sq. ft, making a total of 804.8 sq. ft. The main frames were of iron, and were between the wheels, and the pumps were placed between the frames, the pump rams being worked by eccentrics on the driving axle. A compensating lever was employed for the bogie bearings, having one spring on each side of the bogie. The safety valves were placed over the firebox, and the regulator in the usual position in the boiler. The capacity of the saddle tank was 350 gallons, but a tank fixed under the footplate behind the trailing axle held 256 gallons, giving a total water capacity of 606 gallons. Unfortunately, no record has been kept of the weight of this interesting old engine, which for a number of years worked the Uckfield line. It finished its days piloting at Portsmouth, and was broken up in the late 1870s.

In July, 1859, two small single-wheel express engines were put to work, numbered 137 and 138, *(Fig. 40)*. They had cylinders 15 inch diameter by 22 inch stroke, the centres being placed 2 feet 5 inches apart; the diameter of the blast pipe orifice was 4¼ inches. The leading and trailing wheels were 4 feet in diameter, and the driving 6 feet 3½ inches. The wheelbase was 14 feet 6½ inches, the distance from leading to driving centres being 7 feet, and from driving to trailing 7 feet 6½ inches. The external diameter of the boiler barrel was 3 feet 9 inches, and its length 11 feet 4½ inches; it was constructed in three rings, and the plates were 7/16 inch thick, but at the circumferential joints they were thickened up to ½ inch, which must have made it a rather expensive operation to roll the plates, with apparently little advantage obtained. The centre line of the boiler was 6 feet 2 inches from the rail level. The usual practice of fastening the smokebox to the boiler was, and is also at the present time with most locomotives, by means of a flanged tubeplate and an angle iron ring, but these two engines had the barrel of the boiler flanged outwards, so as to form an angle iron ring, and the smokebox wrapper plate flanged inwards. The front plate of the smokebox was also turned in, as is now frequently done. Each boiler contained 150 brass tubes 2 inches in diameter, and 11 feet 7 inches long. The firebox was 4 feet 5 inches in length outside, and 5 feet 1½ inches deep inside. The heating surface of the tubes was 807 sq. ft, and of the firebox 78 sq. ft,

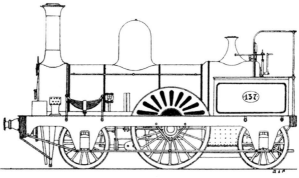

Fig. 40.

making a total of 885 sq. ft, whilst the grate area was 13½ sq. ft. These engines had two spring balance safety valves over the firebox, and the regulator was in the smokebox, which was the general position on most of the engines on the Brighton line up to about this date. An apparatus was fitted in the smokebox of Nos 137 and 138 for controlling the action of the blast on the fire, the chimney being carried down into the smokebox below the top of the blast pipe, on which was a disc capable of being raised towards or lowered from the base of the chimney extension, which was bell-mouthed, thus closing or opening it. This arrangement was connected by a crank to a lever running along the right-hand side of the boiler, and operated from the footplate. A device was also fitted to these engines for obtaining a more perfect combustion of the gases in the firebox. This consisted of a steam jet admitted from the front of the firebox, and an air jet from the back, whilst across the firebox there was a cast-iron baffle plate, similar to the brick arches now in use.

Neither of these arrangements seems to have given very much satisfaction, as it does not appear that any other engines were fitted up in a similar manner. These engines weighed 27 tons 16 cwt each, the leading wheels carrying 10 tons 13 cwt, the driving 10 tons 15 cwt, and the trailing 6 tons 8 cwt. As will be seen from the illustration, they had inside boxes to the driving, and outside to the leading and trailing wheels. The weather board was turned over the full length of the footplate, and was supported by iron pillars. In November and December, 1878, they were re-numbered 382 and 383 respectively, and were both scrapped in January, 1884, after twenty-five years' work, during which, especially in their earlier days, they ran some of the fastest and heaviest trains.

In November, 1859, a small single express engine was built for running light traffic, numbered 108 *(Fig. 41)*, of which the following are the chief particulars: Diameter of cylinders 15 inches, stroke 20 inches, centres 2 feet 7½ inches, steam ports 1¼ inches wide by 10¾ inches long, exhaust ports 4 inches wide, diameter of leading wheels 3 feet 6 inches, of driving 5 feet 8 inches, and of trailing 3 feet, wheel-base 13 feet 2 inches, being 6 feet 0½ inch between leading and driving centres, and 7 feet 1½ inches between driving and trailing. The boiler was made in the usual form, with three rings, its external diameter being 3 feet 6⅜ inches, and its length 10 feet, and containing 147 brass tubes of 1⅞ inch diameter. The steam dome was on the second ring of the barrel, a spring balance safety valve being placed upon it, and another over the firebox. The

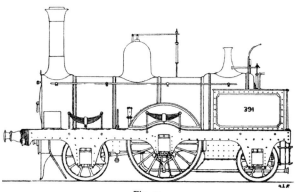

Fig. 41.

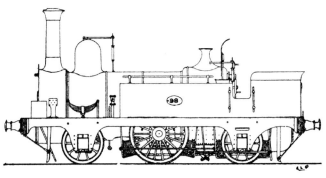

Fig. 42.

centre line of the boiler was 5 feet 10 inches from the rail level, and the length of the firebox was 3 feet 7⅞ inches outside, and its depth 4 feet 7 inches inside. The tubes contained 780 sq. ft of heating surface, and the firebox 67 sq. ft, giving a total of 847 sq. ft. The outside framings of this engine were of the sandwich pattern, and the inside frames were of iron, the latter being fixed between the back of the cylinders and the front plate of the firebox, to which they were attached by cast-iron brackets. This engine only weighed 23 tons 3 cwt, having 7 tons 16 cwt on the leading, 10 tons 10 cwt on the driving, and 4 tons 17 cwt on the trailing wheels. It was re-numbered later 285, and again in June, 1879, 391, being scrapped in January, 1884.

In December, 1859, Craven constructed at Brighton a single-wheel side tank engine for working branch line trains. It was numbered 98, *(Fig. 42)*. The cylinders of this locomotive were 15 inch diameter by 20 inch stroke, and their centres were 2 feet 7½ inches apart. The steam ports were 1¼ inches by 10¾ inches, and the exhaust ports 4 inches by 10¾ inches, the travel of the slide valves being 4½ inches, the lap 1 inch and the lead in full gear 3/16 inch. The leading and trailing wheels were 3 foot 6 inch diameter and the driving wheels 5 feet 6 inches, whilst the wheel-base was 13 feet 6 inches, being 6 feet 6 inches from leading to driving centres and 7 feet from driving to trailing. The external diameter of the boiler, which was made in three rings, was 3 feet 7¼ inches; the length of the barrel was 10 feet and the centre line was 5 feet 8½ inches from the rail level; the boiler contained 147 brass tubes of 2 inch diameter and 10 feet 6 inches long. The length of the outside firebox was only 3 feet 8 inches and the depth of the inside box was 4 feet 7 inches. The steam dome was placed on the first ring of the barrel, with one safety valve fixed thereon and two over the firebox. This engine had outside sandwich framings and inside iron frames between the cylinders and the front plate of the firebox. The side tanks combined contained 260 gallons and a tank under the footplate behind the trailing wheels 250 gallons, giving a total of 510 gallons. The weight of this engine was 28 tons 4 cwt, the leading wheels carrying 9 tons 4 cwt, the driving 10 tons and the trailing 9 tons. Several years ago this engine suddenly left the road whilst running a train on the Lewes and Newhaven branch, owing, it is believed, to alteration of the weight at the trailing end. On another occasion No. 98 was standing in the running shed getting up steam, and the regulator handle was either opened or had been left open previous to lighting up. Fortunately it was hemmed in before and behind by other engines, but it soon accumulated sufficient steam to cause the driving wheels to revolve, and then worked away at such a velocity that it created a blast powerful enough to raise the slates off the roof of the shed. Stroudley put his standard cab over the footplate of this engine and named it *Seaford*. It was afterwards re-numbered 298 and again in May, 1878, 214. We believe the last work this engine did was on the Bognor branch, but owing to the continual collapsing of the boiler tubes it was brought to Brighton, and from there it went to scrap.

Following No. 98 came two single driving wheel express locomotives, numbered 139 and 140. Engine No. 139 had cylinders 15 inch diameter by 20 inch stroke and 6 foot driving wheels, while No. 140 had a pair of 16 inch cylinders and 6 foot 6 inch driving wheels; in all other respects they were of the same dimensions, which were as follows: Diameter of leading and trailing wheels 4 feet, wheel-base 14 feet 6½ inches, being 7 feet between the leading and driving wheels and 7 feet 6½ inches from driving to trailing. The boilers were made in three rings, the external diameter being 3 feet 9⅜ inches and the length 11 feet. The plates were thickened up at the joints, as in the case of Nos. 137 and 138, and the centre line of the boiler from the rail level was 6 feet 3½ inches. As will be seen from *Fig. 43*, they had large cast-iron fluted domes, with polished copper tops, on which was mounted a spiral spring lock-up safety valve. No. 139 had a bell-mouthed top to its dome, in which was encased the safety valve, two spring balance valves being also placed over the firebox. The length of the latter was 4 feet 5 inches and the depth of the inside box 5 feet 1½ inches. The regulator was placed in the usual position in the smoke-box. These engines had iron plate frames, both inside and out, the inside frames running from the front of the cylinders to the

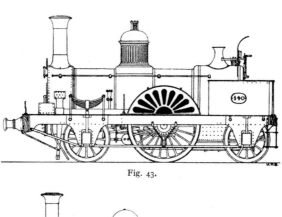

Fig. 43.

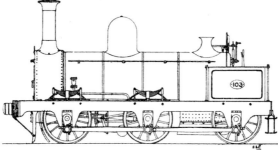

Fig. 44.

front plate of the firebox, to which they were joined by a bracket, the axle bearings for the leading and trailing wheels being outside and those for the driving inside. The weight of these engines was practically the same, but No. 140 was slightly heavier than No. 139, the weight of the former being 28 tons 6 cwt, of which the leading wheels carried 10 tons 8 cwt, the driving 11 tons 8 cwt and the trailing 6 tons 10 cwt. No. 139 was in February, 1879, re-numbered 385, and No. 140 became 386. The former was scrapped in November, 1884, and the latter in September, 1879.

In January, 1860, the company purchased from Slaughter, Grüning & Co. (late Stothert, Slaughter & Co.), of Bristol, two small six-wheels coupled tender engines which that firm had for sale. These were numbered 103 and 114, *(Fig. 44)*. The cylinders were 16 inch diameter, with a stroke of 24 inches. The diameter of the wheels was 4 feet 6 inches, and the wheelbase 15 feet 8 inches, from leading to driving centres being 7 feet 4 inches and from driving to trailing 8 feet 4 inches. The external diameter of the boiler was 4 feet 3 inches and its length 11 feet, the height of the centre line from the rail level being 6 feet 1½ inches. Each boiler contained 191 brass tubes 2 inch diameter by 11 feet 3¾ inches long. The firebox casing was flush with the barrel, instead of being raised, as in most engines then constructed. The length of the outside firebox was 4 feet 10½ inches and the breadth 4 feet 2 inches, the inside depth 5 feet 2 inches. The regulator was in the dome, which was on the second ring of the barrel. There were two spring balance safety valves on the firebox. Inside and outside framings were of solid plate iron, the former stopping at the front of the firebox. These engines weighed about 30 tons 8 cwt each, the leading wheels carrying 10 tons 10 cwt, the driving 13 tons and the trailing 6 tons 18 cwt.

At the commencement of 1860 a four-wheels coupled mixed traffic engine was built at the Brighton Works, which was practically a new locomotive, although composed of some details of two other engines, namely, one of Hackworth's converted coupled engines, which had been re-numbered 114 (originally No. 47), and of No. 89, a Stothert single. This new locomotive was numbered 89, and was the first on the line to be fitted with a Giffard injector. The principal dimensions of this engine were: Cylinders 15 inch diameter by 24 inch stroke, diameter of leading wheels 3 feet 9 inches and of driving and trailing wheels 4 feet 9 inches, centre of leading to centre of driving wheels 6 feet 9 inches, and driving to trailing centres 6 feet 6 inches. The boiler was made in three rings with the steam dome on the second ring, and was 10 feet 6 inches long by 3 foot 8⅞ inch external diameter, and contained 135 brass tubes of 2 inch diameter and 10 foot 11½ inch length, the centre line being 5 feet 8½ inches from the rails. The outside firebox was 4 feet long by 3 feet 7 inches wide, its depth below the boiler centre being 4 feet 0 7/16 inch. One spring balance safety valve was fixed on the dome and another over the firebox. The engine weighed 27 tons 16 cwt, having 9 tons 4 cwt on leading wheels, 11 tons 6 cwt on the driving and 7 tons 6 cwt on the trailing.

In May, 1860, a small outside cylinder single-wheel tank engine was built at Brighton for branch line service. Numbered 25, it was not strictly a new one, for it was built out of parts of a single engine made by Sharp's in 1839, which came into the hands of the company upon the dissolution of the Joint Locomotive Committee in 1846. How much of the original locomotive went into the new one we cannot say, but probably

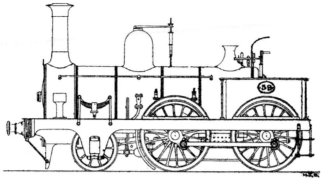

Fig. 45.

only the wheels and framing. Very few particulars are available, but the cylinders were 13 inch diameter by 18 inch stroke; the diameter of the leading and trailing wheels was 3 feet 6 inches and of the driving 5 feet 6 inches. The boiler was 3 feet 6 inches in diameter and 8 feet in length, and contained 132 tubes 1¾ inches in diameter and 8 feet 6 inches long. This engine weighed 29 tons 13 cwt, having 10 tons 17 cwt on the leading, 11 tons on the driving and 7 tons 16 cwt on the trailing wheels. The tank was placed on the top of the boiler. Its number was altered in May, 1875, to 268, and again in May, 1880, to 399. In January, 1884, it was broken up.

During the same month – May, 1860 – there was constructed at Brighton Works a four-wheels coupled tender engine, which was numbered 59, *(Fig. 45)*. This had inside bearings to the driving and trailing wheels, outside bearings to the leading wheels, and a flush-topped firebox. The driving and trailing splashers were open, with double brass bands running round them. The cylinders were 15 inch diameter, and were placed horizontally, the stroke of pistons being 22 inches and the centres placed 2 feet 5 inches apart. The travel of the valves was 4½ inches, lap 1 inch and lead 3/16 inch; steam ports 1 inch by 12 inches and exhaust ports 3½ inches by 12 inches. The leading wheels were 3 foot 6 inch diameter and the driving and trailing wheels 5 feet 6 inches. The wheelbase was 13 feet 7 inches, being from the leading to the driving centres 6 feet 5 inches, and 7 feet 2 inches from the driving to the trailing. The external diameter of the boiler was 3 feet 9¾ inches and its length 10 feet. It contained 137 tubes 2 inch diameter and

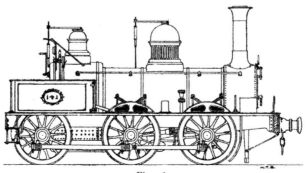

Fig. 46.

10 feet 5½ inches long, its centre line being 6 feet from the rail level. The length of the outside firebox was 4 feet 5 inches, the breadth 3 feet 11 inches and the height inside 4 feet 8½ inches. As mentioned, the firebox casing was made flush with the barrel of the boiler, one spring balance safety valve being fixed on the dome and two over the firebox. The regulator was in the smokebox and the boiler was fed by pumps.

The frames were of iron, but for the leading bearings sandwich framing was adopted. The weight of this engine was 26 tons 16 cwt, the leading wheels carrying 9 tons 16 cwt, the driving 11 tons 4 cwt and the trailing 5 tons 16 cwt. It was afterwards converted into a tank engine by Stroudley, who put in his standard type of boiler, side tanks and cab. After this it was named *Leatherhead*, and in October, 1875, re-numbered 276: in December, 1879, it was again re-numbered 410, and scrapped in May, 1880.

In October, 1860, two six-wheels coupled goods engines were built, Nos 141 and 142. They were rather old-fashioned for that date, their appearance suggesting that they had been built from ten to fifteen years earlier. They had inside framings and two steam domes, one of which was fluted, with a bright copper top, on which was placed a spring balance safety valve; this dome was on the barrel of the boiler, and the other dome, which was not fluted, but of a rather fancy design, was over the firebox, and on it also was a spring balance safety valve. These engines are illustrated by *Fig. 46*, some of their leading dimensions being – diameter of cylinders 16 inches, stroke 24 inches; diameter of wheels 4 feet 9 inches; boiler 3 foot 6 inch diameter and 10 feet 6 inches long, containing 152 tubes 2 inch diameter. The average weight of these two engines was 27 tons 16 cwt each, the leading wheels carrying 10 tons 8 cwt, the driving wheels 11 tons 2 cwt and the trailing wheels 6 tons 6 cwt. No. 141 was scrapped in March, 1878, and No. 142 in November, 1878. Unfortunately no further records have been retained of these two interesting engines.

The next locomotive built at Brighton was a four-wheels coupled leading bogie tank engine, having outside cylinders, and numbered 144; it was constructed in June, 1861, *(Fig. 47)*. This had horizontal cylinders 15 inch diameter, stroke of pistons 22 inches, centres of cylinders apart 6 feet 2 inches; diameter of bogie wheels 3 feet, diameter of driving and trailing 5 feet; wheel-base of bogie 5 feet 11 inches, centre of bogie to centre of driving axle 9 feet 9¾ inches, driving to trailing centres 6 feet 5½ inches, total wheel-base 19 feet 2¾ inches. The engine, instead of taking its bearing on the centre

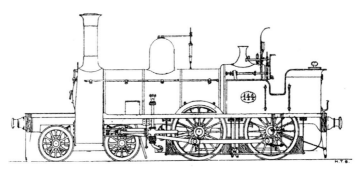

Fig. 47.

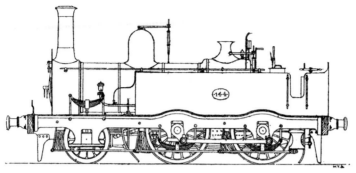

Fig. 48.

of the bogie, was supported on angle iron brackets fixed on the four corners of the engine and bogie framings. The bogie had no lateral movement, moving only around the bogie centre pin. The framing was of plate iron, and had inside bearings for all the wheels. The boiler was built in three rings, and was of 3 foot 9⅝ inch diameter outside, with a length of 10 feet 6 inches, and containing 135 tubes of 2 inch diameter, its centre line being 5 feet 11 inches above the rails. The length of the outside firebox was 3 feet 11½ inches and the height inside 4 feet 8½ inches. The tubes contained 758 sq. ft of heating surface and the firebox 65 sq. ft, giving a total of 823 sq. ft. The regulator was in the smokebox; one spring balance was fixed on the dome and two over the firebox. The pumps for feeding the boiler were placed outside the motion plate; the connecting rods were forked for about half their length, and, as will be seen from the illustration, the smokebox had an inclined front. A water tank was placed under the barrel of the boiler in front of the driving wheels, and another behind the trailing axle, under the footplate. The weight of this engine was 30 tons 4 cwt, distributed as follows: On the leading wheels 10 tons, on the driving 10 tons 8 cwt and on the trailing wheels 9 tons 16 cwt. In July, 1868, No. 144 was converted by Craven into an inside cylinder tank engine, having new outside framing and side tanks. Larger wheels were also adopted and outside bearings. This engine is illustrated in its altered condition by *Fig. 48*. As a matter of fact, about the only thing which remained of the original was the boiler. The cylinders were now 16 inch diameter, stroke 20 inches, distance apart of centres 2 feet 5 inches; diameter of leading wheels 4 feet, diameter of driving and trailing 5 feet 6 inches. The wheel-base was made 15 feet, from leading to driving centres being 7 feet 9 inches and from driving to trailing centres 7 feet 3 inches. The boiler centre was lifted up 3 inches, making its height from the rails 6 feet 2 inches. The side tanks and the tank under the footplate were together capable of carrying 715 gallons of water. The hand brake of this engine actuated blocks on the leading wheels as well as on the driving and trailing wheels, this being rather unusual on the old Brighton engines. After these alterations this locomotive became a few tons heavier, it having 9 tons 12 cwt on the leading wheels, 13 tons 12 cwt on the driving wheels and 11 tons 11 cwt on the trailing wheels, making a total of 34 tons 15 cwt. It was scrapped in March, 1878.

During the same month that No. 144 was built, June, 1861, engine No. 145 left the shops. It was intended for mixed traffic work, and as will be noticed in *Fig. 49,*

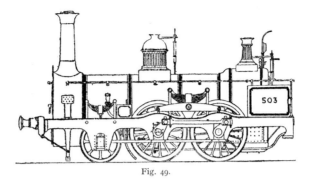

Fig. 49.

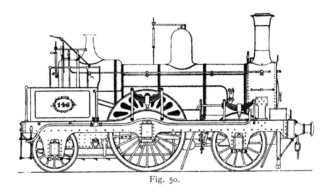

Fig. 50.

all the wheels were placed in front of the firebox; it had a large fluted dome, with a polished copper top, on which was mounted a safety valve, and over the firebox was a fluted safety valve funnel. The dimensions were: Cylinders (horizontal) 15 inch diameter, stroke 22 inches, centres 2 feet 7 inches apart; diameter of leading wheels 4 feet, diameter of driving and trailing wheels 5 feet; leading to driving centres 5 feet 1½ inches, driving to trailing 5 feet 4 inches, total wheel-base 10 feet 5½ inches; from the front of the buffer beam to centre of leading wheels was 6 feet 7 inches, and from the centre of trailing wheels to back of trailing beam 5 feet 6⅜ inches. The boiler was in three rings, the diameter inside the largest ring being 3 feet 6 inches; the length of the barrel was 12 feet 8¼ inches. It contained 125 tubes 2 inch diameter and 13 feet 1 inch long, the regulator being in the smokebox, and the height of the centre line from rail level 5 feet 7 inches. The length of the outside firebox was 3 feet 8½ inches and its width 4 feet, while the length of the inside firebox was 3 feet 2½ inches; its height 4 feet 9 inches and its width 3 feet 4 inches. The heating surface of the tubes was 850 sq. ft, and that of the firebox 65 sq. ft, giving a total of 915 sq. ft. This engine had outside sandwich framings and bearings, and inside bearings also to the driving and trailing wheels; the inside frames were of iron, commencing at the back of the cylinders and finishing at the side of the firebox about on its centre line. The motion plate, like that of many other engines, only extended about half-way across the engine, there being, in fact, two, one for each cylinder, connected together by a stay. The weight of No. 145 was 27 tons, the leading wheels carrying 8 tons, the driving 10 tons and the trailing

9 tons. In October, 1860, it was re-numbered 503, and was scrapped in June, 1882.

In September, 1861, two fine single express engines were built at the company's shops for running the Brighton and London express, and were numbered 146 and 147. These locomotives, illustrated by *Fig. 50*, were considered large at the time they were constructed, and indeed were so compared with the previous engines that Craven had designed. Cylinders were 17 inch diameter by 22 inch stroke, and were placed horizontally. The diameter of the leading wheels was 4 feet 6 inches, of the driving wheels 6 feet 6 inches and of the trailing wheels 4 feet. The wheel-base was 15 feet 10 inches, being 7 feet 11 inches both from leading to driving and from driving to trailing centres. The boilers of these engines were made in three rings, the external diameter being 3 feet 11 inches and the length 11 feet, and contained 158 tubes 2 inches in diameter and 11 feet 6 inches long. The centre line of the barrel was 6 feet 8 inches above the rail level, and the firebox had a length outside of 5 feet 2½ inches, width outside 4 feet 3 inches and height inside 5 feet 5¼ inches. The heating surface of the tubes was 949 sq. ft and of the firebox 122 sq. ft, giving a total of 1,071 sq. ft. From the illustration it will be seen that these engines had very small steam domes, and the railings round the side sheet will also be noticed. Another peculiarity was that they had four sandboxes, one in front of the leading wheels on each side, and also one in front of the driving splashers on each side. The pipe which is seen leading from the smokebox to the trailing end of the engine is for conveying exhaust steam to the tender tank for heating the feed water. This arrangement was very common on the Brighton engines, especially on those which had pumps and no injectors. At this period some of the locomotives had two pumps and one injector, and some one of each, whilst others had pumps only. Nos 146 and 147 had double fireboxes, a partition running from the front to the back. Through this were five longitudinal stays, connecting together the front and back plates of the outside casing. They had sandwich framings and outside bearings to all the wheels, the driving having also inside bearings. The inside framings were of iron, and extended from the leading buffer beam to the front plate of the firebox, to which they were connected by a cast-iron bracket. The regulator was in the smokebox. The average weight of these engines was 31 tons 5 cwt, the leading wheels carrying 10 tons 14 cwt, the driving 12 tons 18 cwt and the trailing 7 tons 13 cwt. During Stroudley's administration No. 146 was named *Lancing* and No. 147 *Worthing*, and, as in the case of all other single wheel engines with outside bearings that were named, had the driving springs underhung, and a new solid combination sandbox and splasher, as well as Stroudley's standard type of bearing springs to all the wheels, and various general details, such as brass number plates, boiler mountings, etc. These engines, like a few others when they were new, had their numbers painted on. No. 146 had a copper chimney cap and No. 147 a brass one. They were latterly fitted with the Westinghouse automatic air brake, and in October, 1880, were re-numbered 451 and 452. Both were scrapped in April, 1886; they were claimed to be at the time the two prettiest engines on the LBSCR.

They were followed by another express engine built in October, 1861, and numbered 84, while in August 1862, another of practically the same type, but with a few minor differences in dimensions, was turned out and numbered 86. They were later re-

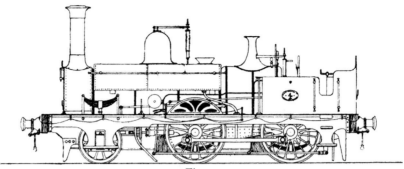

Fig. 51.

numbered 48 and 49. Old numbers 84 and 86, it will be remembered, were constructed by Stothert & Slaughter, and the two new engines taking these numbers had the original framings and wheels of the Stothert engines, but everything else was new, including boilers, cylinders, valve motion, etc., and the weather boards, which latter were not on the old engines in their original condition. Unfortunately, few particulars are available concerning these engines, but the cylinders were 15 inch diameter by 22 inch stroke, diameter of leading and trailing wheels 3 feet 8 inches, and of driving wheels 6 feet; wheel-base 14 feet; diameter of boiler 4 feet 0½ inches, length 10 feet, number of tubes 207. They weighed about 25 tons 18 cwt each, having 8 tons 12 cwt on the leading wheels, 12 tons 6 cwt on the driving, and 5 tons on the trailing.

In January, 1862, a single-wheel saddle tank engine was constructed at the company's works for running branch line trains. This locomotive, numbered 4, had cylinders 13 inch diameter by 22 inch stroke, diameter of leading and trailing wheels 3 feet 6 inches, and of driving wheels 5 feet 11 inches. The leading and trailing wheels had outside bearings, whilst those of the driving wheels were inside. In January, 1869, Craven converted No. 4 into a four-wheels coupled engine, and in order to effect this alteration it was found necessary to add a new framing at the trailing end, and connect it to the original, the old end being of course removed, and thus altered this locomotive is illustrated by *Fig. 51*. New cylinders were put in, which were 16 inch diameter by 20 inch stroke; distance apart of centres 2 feet 5 inches. The diameter of the coupled wheels was made 5 feet, and the wheel-base was 16 feet equally divided. The boiler, which remained the same, was 3 feet 7¼ inches in. diameter, 11 feet long, and contained 104 2 inch tubes. Its centre line was very low, being only 5 feet 9 inches from the rail. The length of the outside firebox was 4 feet 10½ inches, and its breadth 3 feet 9 inches, whilst the inside firebox, which had an inclined grate, was 4 feet 6½ inches deep in front, and 3 feet 8 inches at the back. A spring balance safety valve was fixed on the dome, and two over the firebox. Both the inside and outside framings were of iron, the former running the whole length of the engine, the leading end remaining the same as when it was a single. The driving and trailing springs, which had previously been above the axleboxes, were underhung afterwards; the boiler was fed by one pump and one injector. In addition to the saddle tank there was a small one behind the trailing axle underneath the footplate. The weight of this engine was 29 tons 10 cwt, the leading wheels carrying 9 tons 14 cwt, the driving 11 tons 12 cwt, and the trailing 8 tons 4 cwt. This engine was re-numbered, firstly 104, and then 295. In October, 1877, it was made 365, and broken up in March, 1882.

Following this locomotive came two small coupled passenger engines, which were built for working ordinary main line stopping trains. They were put to work in January, 1862, and numbered 32 and 148, but, unfortunately, little information seems to have been preserved concerning them. The cylinders were 16 inch diameter, stroke 20 inches, diameter of leading wheels 4 feet, diameter of driving and trailing wheels 5 feet 6 inches, wheel-base 13 feet 9 inches, diameter of boiler 3 feet 11¾ inches, length 10 feet 0¾ inch, number of tubes 180, diameter 2 inches. The weight of these engines was about 28 tons 7 cwt each, thus distributed: On leading wheels 9 tons 11 cwt, on driving 12 tons 3 cwt, and on trailing 6 tons 13 cwt. These two engines had outside bearings to all the wheels, and in general design were very much the same as those built by Sharp, Stewart & Co. for the company in 1854–6, with the exception of the position of the steam dome. In March, 1876, No. 32 was re-numbered 275, and in December, 1879, was again altered to 411, being broken up in September, 1882. No. 148 was altered in external appearance by Stroudley, having the driving and trailing springs suspended, and the open driving splashers substituted by closed-in ones. We may remark that when a single-wheel engine was named by Stroudley a combined sandbox and splasher were used, but in the case of coupled engines these were kept separate; of the latter type, however, very few were named. At the time No. 148 was altered it was named *Ryde*; in October, 1880, it was re-numbered 453, and in November, 1886, it went to the scrapheap.

At the beginning of 1862 Craven introduced his standard class of four-wheels coupled express engines, if indeed a standard type can have been said to exist before the year 1871, when Stroudley assumed control of the locomotive department. However, nearly all the coupled express engines built after this date were the same, with the exception of slight differences in dimensions and details, although it must be remarked that hardly anything was interchangeable; the day for that had not arrived. There were twenty-nine engines of this general design, all the various types of which will be noticed in due order. Craven built eleven at Brighton, twelve were by Beyer, Peacock & Co., and six by Dübs & Co. In January, 1862, the first two of these engines were turned out of the company's shops, numbered 149 and 150 *(Fig. 52)*. Their chief

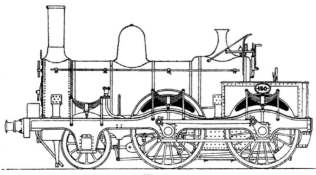

Fig. 52.

dimensions: Diameter of cylinders 16 inches, stroke of pistons 20 inches, centres of cylinders apart 2 feet 5 inches; diameter of blast pipe orifice 4½ inches; diameter of leading wheels 4 feet, diameter of coupled wheels 6 feet; leading to driving centres 6 feet 6 inches, driving to trailing centres 7 feet 3 inches, total wheelbase 13 feet 9 inches. The barrel of the boiler, which was of iron, was constructed in three rings, its external diameter being 4 feet 1 inch, and its length 10 feet 0¾ inches, and in it were 180 brass tubes, 2 inches in diameter and 10 feet 5 inches long. The steam dome, which was on the second ring of the barrel, was 1 foot 11 inches in diameter outside, and on it was a fixed spring balance safety valve, two others being over the firebox. The length of the outside firebox was 4 feet 8 inches, while the inside box was in length 4 feet, in breadth 3 feet 8¼ inches, and in height 5 feet 6 inches; it was of copper, and connected to the firebox casing by girder-roof stay bars, this being a standard practice at that date. The centre line of the boiler was 6 feet 5¼ inches from the rail level. The fire grate area was 14.96 sq. ft and the heating surface 1,064 sq. ft, the tubes giving 974 sq. ft and the firebox 90 sq. ft. The working pressure of the boiler was 140 lbs per sq. in. The slide valves were actuated by the ordinary Stephenson link motion, the weigh-bar shaft being above. The intermediate valve rods were suspended by a link at a point forward of the expansion link, the eccentric rods being connected to lugs forged solid to the links. This valve gear was locally known as 'Craven's'. The regulator was in the smokebox. All the wheels had outside bearings, the driving having inside bearings also. Sandwich framings were used outside, the inside frames being of plate iron, extending from the leading buffer beam to the front of the firebox casing, to which they were connected in the usual manner by a bracket. No. 150 was the first engine on the LBSCR to have a brick arch in the firebox. The weight of these locomotives was about 27 tons 16 cwt each, the leading wheels carrying 9 tons 2 cwt, the driving 12 tons 7 cwt, and the trailing 6 tons 7 cwt. They were fitted by Stroudley with the Westinghouse air brake, and in October, 1880, were re-numbered 454 and 455. The former was scrapped in August, 1884, and 455 in April, 1883.

In February, 1862, two more locos of the type just described were constructed at the Brighton works, numbered 151 and 152. They were exactly the same as Nos 149 and 150 with the exception of the valve motion and a few slight differences in dimensions. *Fig. 53* illustrates these engines, of which the wheel-base was 9 inches longer than in

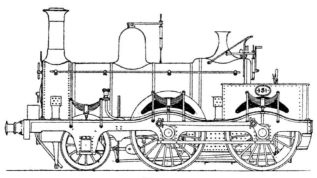

Fig. 53.

the previous two engines, being 6 feet 9 inches from leading to driving centres, and 7 feet 9 inches from driving to trailing, total 14 feet 6 inches. The framings were all of iron, and the inside framings – like those of Nos 149 and 150 – stopped at the front plate of the firebox, to which they were connected, but in Nos 151 and 152 they again commenced at the back of the firebox, and were continued to the end of the footplate. The motion plate, as in a case previously noted, only extended half-way across the engine. The slide valves were actuated by the ordinary Stephenson link motion, the weigh-bar shaft being below, and the intermediate valve rods suspended at the back end by a link instead of passing through a guide. The eccentric rods were connected to the links themselves as in more recent practice. The average weight of these two engines was 30 tons each, there being about 9 tons 7 cwt on the leading, 13 tons 5 cwt on the driving, and 7 tons 8 cwt on the trailing wheels. During Stroudley's *régime* they had the safety valves removed from the firebox and put on the dome, the sandboxes in front of the driving wheels and the smokebox stay being removed. They were also fitted with the Westinghouse air brake, and the open splashers and side sheet of No. 152 were filled in. No. 151 was altered to 120, and was again re-numbered 363 in August, 1877, and in April, 1883, was purchased by the West Lancashire Railway, by whom it was numbered 7, and named *Blackburn*. No. 152 was re-numbered 456 in December, 1880, and was scrapped in November, 1887.

In May, 1862, two four-wheels coupled 'long boiler' engines for mixed traffic, were constructed at the Brighton Works. These were numbered 45 and 99, and were the last of this type on the LBSCR. They had cylinders placed horizontally, 15 inch diameter by 22 inch stroke, distance apart of centres 2 feet 7 inches. The wheel-base was 10 feet 5½ inches, from leading to driving centres being 5 feet 1½ inches, and from driving to trailing 5 feet 4 inches. The leading wheels were 3 feet 6 inches in diameter, and the driving and trailing wheels 5 feet. The boilers were made in three rings, the largest being in the middle, which was a standard practice of Craven's; the external diameter was 3 feet 8 inches, and the length of the barrel 12 feet 8¼ inches. They each contained 158 tubes 2 inch diameter and 13 feet 1¾ inches long, and the height of the centre line from the rail level was 5 feet 9⅝ inches. As will be seen from *Fig. 54*, the steam dome, which was 1 foot 11 inches in diameter, was placed in the middle of the boiler. The length of the outside firebox was 4 feet 2¼ inches, breadth 3 feet 11¾ inches, height of inside

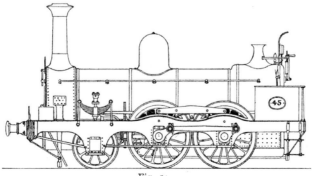

Fig. 54.

firebox 4 feet 10 inches. The roof of the inside firebox was secured to the boiler casing by the ordinary girder roof stay bars. The heating surface of the tubes was 1,096 sq. ft, and of the firebox 71 sq. ft, giving a total of 1,167 sq. ft. These engines had outside sandwich framings and inside iron frames from the back of the cylinders to the side of the firebox casing. To the driving and trailing wheels was fixed a compensating lever and inverted laminated spring which actuated on both wheels. The regulator was in the smokebox, but the steam was conveyed to it by a pipe led up into the dome. The boiler was fed by two pumps worked off the crossheads. The motion plate only extended half-way across the engine as in others previously mentioned. The leading wheels had outside bearings only, whilst the coupled wheels had both inside and outside. The slide valves were worked by the Craven variety of the Stephenson link motion described previously. The average weight of these engines was about 29 tons 6 cwt each, the leading wheels carrying 6 tons 14 cwt, the driving 11 tons 4 cwt, and the trailing 11 tons 8 cwt. In June, 1877, No. 45 was re-numbered 277, in December, 1879, was again re-numbered 409, and in May, 1880, was broken up.

Following these engines came two very handsome singles resembling the 'Jenny Lind' type, which were numbered 153 and 154, and built at Brighton in July, 1862, *(Fig. 55)*. Their chief particulars: Cylinders, placed horizontally, 16 inch diameter, stroke of pistons 20 inches, centres of cylinders apart 2 feet 5 inches; diameter of leading and trailing wheels 4 feet 3 inches, diameter of driving 6 feet 6 inches, total wheel-base 14 feet 7 inches, from leading to driving and driving to trailing being each 7 feet 3½ inches. The boiler was made in three rings, the largest external diameter being 4 feet 0½ inch, and the length of the barrel 10 feet 0¾ inch. It contained 180 brass tubes 2 inch diameter and 10 feet 6 inches long. The steam dome was on the second ring of the boiler, and on it was fixed a spring balance safety valve, another being over the firebox. The centre line of the boiler was 6 feet 8 inches from the rail level. The length of the outside firebox was 4 feet 8 inches, the breadth 4 feet 3¼ inches, and the depth 5 feet 8 inches. The heating surface of the tubes was 990 sq. ft, and of the firebox 92 sq. ft, giving a total of 1,082 sq. ft. The boiler was fed by pumps, the regulator was in the smokebox as in the two engines last described, and the valve motion was the ordinary Stephenson link having the weigh-bar shaft below. The leading and trailing wheels had outside boxes, and the driving inside; all the springs were above the axle boxes. The

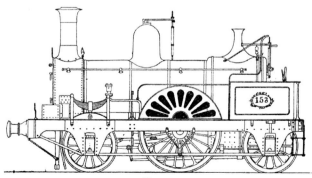

Fig. 55.

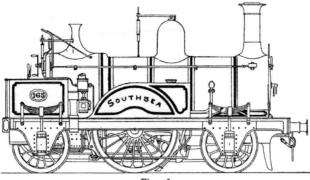

Fig. 56.

outside framings were of iron and likewise the inside, which latter extended from the front of the cylinders to the front plate of the firebox where they were connected up in the usual manner. The weight of these engines was about 28 tons 12 cwt, 10 tons 4 cwt being on the leading, 11 tons 12 cwt on the driving, and 6 tons 16 cwt on the trailing wheels. When Stroudley came to Brighton he made several alterations to these engines, as may be seen in *Fig. 56*, which shows them in their later state. The open splashers were done away with and the standard combination sandbox and splasher adopted, and also new inside framings and spiral springs to the driving wheels. A steel leading buffer beam replaced the original wooden one, whilst a wooden one was substituted for the original iron trailing beam. Shorter side sheets took the place of the old ones and new footsteps were added, as well as bearing springs and various minor details; cylinder lubricators being fixed on the front of the smokebox, and the engines named *Spithead* and *Southsea* respectively. They were latterly fitted with the Westinghouse air brake. No. 153 had the safety valve removed from the firebox and put on the dome, but No. 154 ran to the last with it in the original position. In October, 1876, these engines were re-numbered 164 and 165. Their last runs were on the Tunbridge Wells and Brighton line, and in October, 1890, No. 165 was sent to the scrap heap, to be followed by No. 164 in December, 1891, after 29 years of constant hard work, for it should be remarked before we close the history of these two engines, that they did some of the very best work that a Brighton locomotive was called upon to do. They were originally built for the Portsmouth and London traffic.

In August, 1862, Craven introduced an entirely new class of engine so far as the Brighton Company was concerned, the locomotives in question having the leading and driving wheels coupled. Numbered 155 and 156, *(Fig. 57)*, they were built for working either goods or passenger trains as might be required. The cylinders, which were placed at an inclination of 1 in 9 and 2 feet 5 inches apart, were 16 inch diameter by 20 inch stroke. The leading and driving wheels were 5 foot 6 inch diameter on tread, and the trailing wheels 4 feet. The wheel-base was very short, the total being only 13 feet 3 inches, from leading to driving 6 feet 6 inches, and from driving to trailing 6 feet 9 inches. The boilers were made in three rings with the steam dome in the middle, and the regulator was of the double-beat type similar to that introduced by Ramsbottom on the LNWR. The external diameter of the boiler was 3 feet 10⅞ inches, its length 10 feet, and it contained 182 2-inch diameter brass tubes; the height of its centre line above the rail level was 6 feet 6 inches. One spring balance safety valve was on the dome, and two over the firebox. The regulator handle was hooked at the end, which we believe was the only instance of the kind on this railway. The length of the outside firebox was 4 feet 1 inch, the breadth 3 feet 11 inches, and the height inside 5 feet 1 inch. The boiler was fed by two pumps, and, as will be seen from the illustration, all the axle bearings were inside the wheels. The framing was of iron, and the link motion was of the Stephenson type, having the weigh-bar shaft below. The engine weighed about 26 tons 14 cwt each, distributed as follows: On leading wheels 10 tons 9 cwt, on driving 10 tons 12 cwt, and on trailing 5 tons 13 cwt. In December, 1880, No. 155 was re-numbered 504, and in June, 1882, was scrapped. No. 156 was never re-numbered, but in 1878 was converted into a stationary engine for working the winding apparatus used for hauling wagons up the incline from the wharf to the railway at Kingston-on-Sea, a goods station situated between Shoreham and Southwick. We may add that the engine has long since been replaced by one built for this purpose.

In December, 1862, an engine was built for express work, No. 31, *(Fig. 58)*. This had inside driving bearings and outside for the leading and trailing axles. The cylinders, which were placed horizontally, had a diameter of 15 inches, and a stroke of 22 inches, the centres being 2 feet 7 inches apart. The leading and trailing wheels were 4 feet in diameter and the driving 6 feet 6 inches. Wheelbase was 14 feet 3 inches, being 7 feet 1½ inches between each pair of wheels. The boiler was of the usual design, the external diameter of the largest ring being 4 feet 1 inch and the length 10 feet; it contained 167 tubes, 2 inch diameter and 10 feet 5 inches long, the centre line being 6 feet 7 inches above the rail level. Length of the outside firebox was 4 feet 3 inches, the breadth 4 feet 1 inch, and the height inside 5 feet 1 inch. The heating surface of the tubes was 911 sq. ft, and of the firebox 80 sq. ft, giving a total of 991 sq. ft. The weight of this engine was 28 tons 13 cwt, 9 tons 16 cwt being on the leading, 12 tons 18 cwt on the driving, and 5 tons 19 cwt on the trailing wheels. A double-beat regulator was in the dome, which also carried a spring balance safety valve, another being over the firebox. The valve motion was of the Stephenson type, and the outside and inside frames, which were of iron, extended the whole length of the engine. During Stroudley's time it was named *Littlehampton*, at which place it was stationed, and was fitted with the Westinghouse

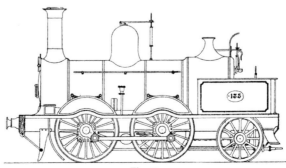

Fig. 57.

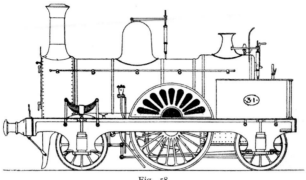

Fig. 58.

air brake. In 1879 it was re-numbered 259, and in May, 1880, was broken up.

During the same month that No. 31 was built, two six-coupled goods engines were also constructed, numbered 157 and 158. These had cylinders 16 inch diameter by 24 inch stroke, diameter of wheels 5 feet, wheel-base 15 feet 9 inches, diameter of boiler 4 feet 3 inches, length 11 feet, number of tubes 200, 2 inch diameter and 11 feet 5 inches long. Owing to the great width of the firebox casing, the tyres of the trailing wheels had to be turned down so thin that the engines very often got off the road. The engines weighed about 34 tons 3 cwt each, the leading wheels carrying 11 tons 10 cwt, the driving 13 tons 15 cwt, and the trailing 8 tons 18 cwt. They had iron framings and outside bearings. The regulator handles pulled out in the same manner as those used on the South-Eastern and Great Northern railways. They were never re-numbered, and we are unaware as to when they were taken out of service. The boiler of 157 was eventually used for supplying steam to a stationary engine.

Engines Nos 159 and 160, *(Fig. 59)*, were built at Brighton in March, 1863, and were four-wheels coupled express locomotives. The cylinders were horizontal and had a diameter of 16 inches and a stroke of 20 inches, the centres being 2 feet 5 inches apart. The diameter of the blast pipe orifice was 4½ inches. The leading wheels were 4 feet in diameter, and the driving and trailing 6 feet. The wheelbase was 15 feet 1 inch, being 7 feet 4 inches from leading to driving centres, and 7 feet 9 inches from driving to trailing. From the leading wheel centre to the front buffer beam was 5 feet 0½ inch,

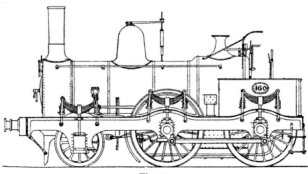

Fig. 59.

and from the trailing centre to the end of the framing was 3 feet 7 inches. The boilers were made in three rings, the largest external diameter being 4 feet 0¾ inches, the length of the barrel 10 feet 0¾ inches, and the height of the centre line above rail level 6 feet 0¼ inches. Ten longitudinal stays passed through the boiler connecting the front tube plate and the back plate of the firebox casing. The length of the outside firebox was 4 feet 8 inches, breadth 4 feet 3¼ inches, and height inside 4 feet 8 inches. The dome was on the middle ring of the boiler and was 1 foot 11 inches in diameter, and 2 feet 9 inches in height, and on it was a spring balance safety valve, two more being over the firebox. The boiler contained 182 brass tubes 2 inch diameter and 10 feet 6 inches long. The heating surface of the tubes was 910 sq. ft, and of the firebox 94 sq. ft, making a total of 1,004 sq. ft. With this heating surface, and the fine proportions of the boiler and firebox, these engines were able to steam well, and were notably good engines. They originally had double fireboxes, but the midfeather was afterwards removed. The height of the chimney from the rail level was 13 feet, diameter at top 1 foot 5 inches, and at bottom 1 foot 3 inches. These locos had double-beat regulators in the domes, and 5 inch diameter steam pipes in the boilers. The latter were fed by two pumps each. The engines had both outside and inside bearings to the driving wheels, the framings being of iron, the inside plates extended from the front buffer beam to the front plate of the firebox casing, and from the back plate to the end of the engine. The slide valves were actuated by Stephenson's link motion, having the weigh-bar shaft above. The cranks were hooped. The bearing spring hangers of these locomotives were formed of stirrup links with a hole drilled through the bottom, through which the hanger itself passed, and on this were threaded indiarubber discs, on the top of which was a washer and adjusting nuts. The weight of these engines was on an average 30 tons 18 cwt each, the leading wheels carrying 9 tons 12 cwt, the driving 13 tons 6 cwt, and the trailing 8 tons. Never re-numbered they were both scrapped in June, 1884.

In September, 1863, a small single express engine was built for working some of the Portsmouth and London traffic. This was numbered 161, *(Fig. 60)*. The cylinders were 16 inch diameter by 20 inch stroke, and were placed horizontally with the centres 2 feet 5 inches apart. The leading and trailing wheels were 4 feet in diameter and the driving 6 feet. The wheelbase was 13 feet 8½ inches, being 6 feet 11 inches from leading to driving centres, and 6 feet 9½ inches from driving to trailing. The boiler, which was

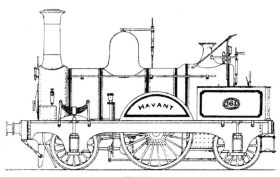

Fig. 60.

constructed according to the usual Brighton practice, was 3 feet 11 inches in diameter and 9 feet 8 inches in length, and contained 162 brass tubes 2 inch diameter and 10 feet 2 inches long; the centre line of the barrel was 6 feet 2 inches from the rails. The length of the outside firebox was 4 feet 2 inches, breadth 4 feet 3 inches, and height inside 5 feet 0½ inches. The steam dome was 2 feet 8½ inches high and 1 foot 11 inches in diameter. This engine had a double-beat regulator, one spring balance safety valve on the dome and two over the firebox. The boiler was fed by two pumps. The heating surface of the tubes was 880 sq. ft, and of the firebox 80 sq. ft, making a total of 960 sq. ft. The valve motion was of the ordinary Stephenson link type, having the weigh-bar shaft below, and the slide valves having a travel of 5 inches. As will be observed from the illustration the driving wheels had inside bearings only, whilst the leading and trailing wheels had outside bearings. The inside and outside framings were of iron, the former stopping at the front of the firebox, which was thus enabled to be of greater width. This engine only weighed 27 tons 5 cwt, having 9 tons 18 cwt on the leading wheels, 11 tons 1 cwt on the driving, and 6 tons 6 cwt on the trailing. During Stroudley's term of office No. 161 was named *Havant*, and fitted with the Westinghouse air brake. It was never renumbered, and in January, 1885, was sent to Horley to be scrapped. This engine, like some of the other singles we have described, did some important work considering its power and size. On several occasions we have known it to run the 4 p.m. fast train from London Bridge to Portsmouth with eight to ten coaches and arrive well to time. No. 161 would, without any difficulty, keep time up the long bank from Horsham to Ockley, which has a gradient of 1 in 90.

We now come to a very important and interesting period of the locomotive history of the LBSCR, for in November, 1863, two large express engines were constructed at the Brighton works, which were probably better known than any other locomotives south of London. They were numbered 162 and 163, and are undoubtedly classed among the finest engines built at Brighton, the good proportions and curves of the framing giving them a striking appearance. Unlike the majority of Craven's engines, all the external details were very well proportioned, the large carrying and driving wheels being in harmony with each other. The safety valve casing was not like the ordinary large and flat one usually adopted on the Brighton engines, but was very similar in appearance to those used on GWR locomotives. They had double brass bands round the driving splashers, and to these the sandboxes were connected, these engines being the first to have them so. They were built expressly for running the 8.45 a.m. Brighton to London express, and the 5 p.m. return train; but were perhaps better known as the *London* and the *Brighton*, which names were given them by Stroudley. They are illustrated, in their original condition, by *Fig. 61*. The following are the chief particulars of these two engines: Diameter of cylinders 17 inches, stroke of pistons 22 inches, placed horizontally with centres 2 feet 6 inches apart. The steam ports were 1½ inches wide, exhaust ports 4½ inches wide and 1 foot 3 inches long. The leading and trailing wheels were 4 feet 6 inches in diameter, and the driving wheels 7 feet, this being the largest diameter of wheels ever used on the LBSCR. The wheelbase was 16 feet 3 inches being 8 feet 0¾ inch from leading to driving centres and 8 feet 2¼ inches from driving to trailing; from the leading wheel centre to the front of the leading buffer beam was

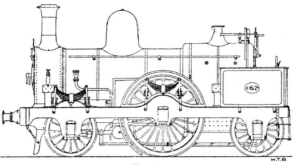

Fig. 61.

4 feet 6½ inches, and from the trailing centre to end of framing 2 feet 8½ inches. The boilers were made as usual in three rings and were 4 feet in external diameter and 10 feet 4 inches in length, the centre line being 7 feet from the rails; they each contained 192 tubes 2 inch diameter and 10 feet 9½ inches long. The steam dome was 3 feet high and 2 feet in diameter. The front tube plate was connected to the firebox back plate by 11 longitudinal stays. A very important feature about these engines were the large double fireboxes, of which the length was 5 feet 11 inches, breadth 4 feet 3½ inches, and height inside 5 feet 11 inches. Between the two fireboxes, in the midfeather, were 5 longitudinal stays, which connected the front and back plates. The tubes contained 1,082 sq. ft of heating surface and the firebox 156 sq. ft, giving a total of 1,238 sq. ft. The boilers were fed by two pumps each, the pump rams being connected to the gudgeons and placed between the crossheads and slide bars. They had double-beat regulators and two spring balance safety valves were over the firebox. Stephenson's link motion was used, the weigh-bar shaft being underneath. The same arrangement of spring hangers as on Nos 159 and 160 was also adopted on these engines, and their average weight was 32 tons 11 cwt, distributed as follows: On the leading wheels 10 tons 11 cwt, on the driving 14 tons, and on the trailing 8 tons. As previously mentioned, they were afterwards named *London* and *Brighton*, and at a later period were re-named *Penge* and *Sandown*, these alterations taking effect when Stroudley's No. 40 *Brighton* and No. 339 *London*, were built in 1878 and 1882 respectively. They were also fitted up with the air brake and other standard details that were introduced since 1871, and are illustrated in their later state by *Fig. 62*. They were never re-numbered, No. 162 being scrapped in January, 1885, and No. 163 in May, 1888.

Following these, two mixed traffic engines were built at Brighton in November, 1863, which were numbered 164 and 165, *(Fig. 63)*. They were designed for the Crystal Palace traffic, No. 164 being stationed at New Cross and 165 at Battersea. The cylinders were 16 inch diameter, stroke of pistons 20 inches. The wheelbase was only 13 feet 6 inches, the coupled centres being 6 feet 3 inches apart, and from driving to trailing centres 7 feet 3 inches. The coupled wheels were 5 feet in diameter, and the trailing pair 4 feet. The boiler was 3 feet 10 inches in diameter, with a length of barrel of 9 feet 9 inches, the centre line being 6 feet 3½ inches above the rails, and it was flush with the firebox casing, this being a very unusual thing in Craven's locomotives.

The boiler had 158 tubes each 2 inches in diameter; heating surface of tubes, 777 sq. ft, firebox 81 sq. ft, total 858 sq. ft. As will be seen from the illustration, these two engines had inside bearings to the coupled wheels, and outside to the trailing. They had double brass bands round the open splashers, and spring balances over the firebox, and had Stephenson's link motion with the weigh-bar shaft below. During Stroudley's time they were chiefly employed on goods traffic. No. 165 was for a time painted the standard colour of Stroudley's passenger engines and used for shunting in the Brighton works yards, being stationed in the works. This painting was afterwards abandoned for the usual goods engine colour. These engines weighed about 26 tons 14 cwt each, the leading wheels carrying 8 tons 17 cwt, the driving 11 tons 5 cwt and the trailing 6 tons 12 cwt. No. 164 was first re-numbered 299, and in May, 1878, 368, being put out of service in August, 1884. No. 165 was altered to 297, and in October, 1877, was

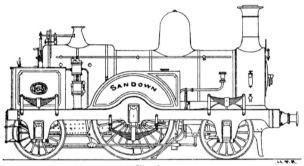

Fig. 62.

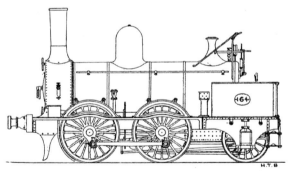

Fig. 63.

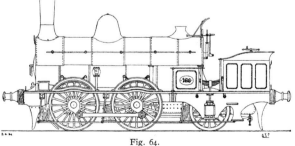

Fig. 64.

re-numbered 364, and broken up in 1887.

In November, 1863, two four-wheels coupled saddle tank engines were built at Brighton for working on the South London line. They were numbered 166 and 167, the former being stationed at New Cross and No. 167 at Battersea, *(Fig. 64)*. Their dimensions were: Diameter of cylinders 16 inches, stroke of pistons 20 inches, centres of cylinders apart 2 feet 5 inches, wheel-base 13 feet 6 inches, from leading to driving centres being 6 feet 3 inches and from driving to trailing 7 feet 3 inches. The leading and driving wheels were 5 feet in diameter and the trailing 4 feet. The boiler was 3 feet 10 inches in diameter and 9 feet 9 inches long, containing 158 tubes of 2 inch diameter, the centre line of the barrel being 6 feet 3½ inches above the rail level. There were no safety valves on the dome, but two spring balances instead over the firebox. The heating surface of the tubes was 777 sq. ft, and of the firebox 81 sq. ft, making a total of 858 sq. ft. These two engines had inside framings for the leading and driving wheels and outside framing for the trailing wheels. The united capacity of the saddle tank and the tank under the coke bunker was 818 gallons, the length of the former being 16 feet 9 inches, and the length of engine over buffer beam 25 feet 5 inches. These engines weighed about 34 tons 15 cwt each, distributed as follows: On leading wheels 12 tons 1 cwt, on driving 12 tons 2 cwt and on trailing 10 tons 12 cwt. They were never re-numbered, No. 166 being broken up in July, 1882, and No. 167 in June, 1877.

In the same month that Nos 166 and 167 were built came two large goods engines, Nos 168 and 169, *(Fig. 65)*. The cylinders were 16 inch diameter by 24 inch stroke, and were placed 2 feet 7 inches apart. The diameter of the wheels was 5 feet and the wheel-base was 15 feet 9 inches, being 7 feet 6 inches from leading to driving centres and 8 feet 3 inches from driving to trailing. From the buffer beam to the centre of the leading wheels was 5 feet 4 inches, and from the centre of the trailing wheels to the end of the framings 3 feet 0½ inch. The boilers were made in three rings, the external diameter being 4 feet 4 inches and the length 11 feet and the centre line of the barrel was 6 feet 6 inches above the rail level. They each contained 200 tubes of 2 inch diameter and 11 feet 5 inches in length. The heating surface of the tubes was 1,187 sq. ft and of the firebox 86 sq. ft, giving a total of 1,273 sq. ft. The length of the outside firebox was 4 feet 10⅛ inches, breadth 4 feet 3⅝ inches and height inside 5 feet 3¾ inches. The boilers were fed by two long-stroke pumps. One spring balance was

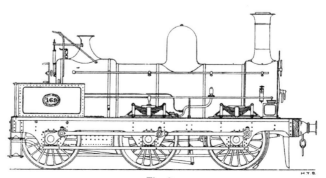

Fig. 65.

fixed on the dome and two over the firebox; but Stroudley removed the one off the dome. Subsequently No. 168 was fitted with injectors, and had the pumps dispensed with. The balances were then taken off the firebox and put on the dome, the old safety valve seating being used for the injector steam cocks. The inside and outside framings were of iron, the former stopping at the front plate of the firebox. The slide valves were controlled by the Stephenson link motion. These engines weighed 33 tons 16 cwt, having 11 tons on the leading wheels, 14 tons 8 cwt on the driving wheels and 8 tons 8 cwt on the trailing wheels. When new they had copper top chimneys, but No. 168, when fitted up with injectors, had a new chimney similar to that used on the LSWR No. 168 was re-numbered 477 in 1894, and scrapped in September, 1896. No. 169 was never renumbered, but scrapped in June, 1893.

In December, 1863, two four-wheels coupled side tank engines were constructed at the Brighton works, numbered 170 and 171. They had some special points of interest about them, having flush-top fireboxes, inside bearings to the coupled wheels and leading wheels in line with the centre of the smokebox, the axle passing under the cylinders, *(Fig. 66)*. The cylinders were 16 inch diameter, and were placed at an angle of 1 in 18¼, stroke of pistons 20 inches, centres of cylinders apart 2 feet 5 inches. The leading wheels were 4 feet in diameter, and the driving and trailing 5 feet. They had a very long wheel-base, namely, 16 feet 8 inches, being 8 feet 11 inches from the centre of the leading wheels to the centre of the driving and 7 feet 9 inches from the driving to the trailing. From the buffer beam to the centre of the leading wheels was 3 feet 5 inches, and from the centre of the trailing wheels to the trailing end buffer beam 5 feet 6 inches. The boilers were in three rings, the external diameter being 4 feet and the length 9 feet 9 inches. They contained 148 tubes 2 inch diameter and 10 feet 1 inch long. The centre line of the barrel was 6 feet 3½ inches above the rail. As mentioned, the outside firebox was made flush with the boiler barrel; its length was 4 feet 8 inches, breadth 3 feet 10 inches and height inside 3 feet 1½ inches. The length of the side tanks was 11 feet, they were capable of carrying 420 gallons of water. In 1869 a small tank was added and put under the framing behind the trailing axle, and carried 250 gallons, thus making a total of 670 gallons. The framings were of iron, excepting the outside framings for the leading wheels, which were of the sandwich pattern. Their weight was 33 tons, the leading wheels carrying 9 tons, the driving 12 tons 2 cwt and the trailing

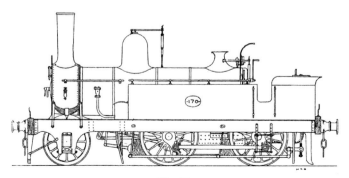

Fig. 66.

11 tons 18 cwt. They were never re-numbered, and in April, 1879, No. 170 was broken up. We have no record of when 171 was scrapped, but believe it was about the same time as No. 170.

These engines were followed by two fine express locomotives, turned out in January, 1864. Numbered 172 and 173, they are illustrated by *Figs. 67 and 68*. The cylinders were 16½ inch diameter, with a stroke of 22 inches, the steam ports being 15 inches long by 1½ inches wide, and the exhaust ports 3½ inches wide. The leading and trailing wheels were 4 feet in diameter and the driving wheels 6 feet 6 inches. The wheelbase was 16 feet, being 8 feet between both pairs of centres. The boiler of this class was 11 feet 6 inches long in the barrel and 3 feet 10 inches in diameter, made of ½ inch iron plate, double riveted in all the seams, with a firebox 4 feet 8½ inches long, by 4 feet 2¼ inches wide and 5 feet 4½ inches high inside. The centre line was 6 feet 6 inches from the rail level, and the heating surface was made up as follows: 153 tubes, 11 feet 9 inches long by 2 inch diameter, 940 sq. ft; firebox, 91 sq. ft, total 1,031 sq. ft, grate area 14.75 sq. ft. The boiler was provided with three safety valves, of which two were placed upon the crown of the firebox and one on the top of the dome; it was fed by two pumps and had no injectors. The regulator was of the double-beat valve type. The framing was of plate iron, inside and outside, the inside frames extending from the buffer plank to the front plate of the firebox, to which they were fixed by wrought-iron brackets; and the outside frames running from end to end of the engine

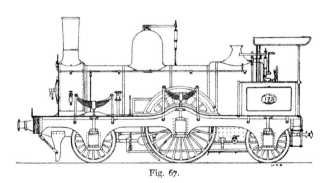

Fig. 67.

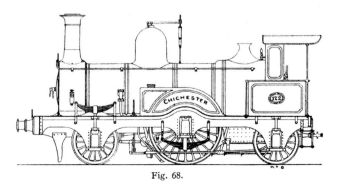

Fig. 68.

and provided with double horn plates, riveted to the main plates. The slide valves were worked by the Stephenson link motion, having the weigh-bar shaft below. The weight of the engines in working order was 31 tons 12 cwt, distributed as follows: On the leading wheels 10 tons 14 cwt, driving 13 tons 10 cwt and trailing 7 tons 8 cwt. The tenders had a water capacity of 1,670 gallons and about 3 tons of fuel, and weighed in working order about 21 tons. These engines ran the express traffic between Hastings and London, and the following are some particulars of a week's working:

No. of Engine	Miles run	Coal burnt	lbs per mile
172	524½	9 tons 7 cwt	39.9
173	813	12 tons 14 cwt	47.3

It will be noticed that the consumption of fuel appears high, which is accounted for by the fact that no allowance was made in the mileage for shunting, forming trains, stoppage and lighting up. From the illustrations it will be seen that these engines had cabs, being the first locomotives on this line to be so fitted; as originally built there were no sides to them, but these were added by Stroudley to No. 172 when he altered it and named it *Chichester*, as shown by *Fig. 68*. No. 173 was never named, but was the first engine to be painted the standard yellow colour. When the engine came out of the shops with that colour the spokes of the wheels were lined out with fine white lines. It was then taken up to London for the directors to see, the result being that Stroudley's colour was adopted, but the lining out of the wheels was not continued owing to the expense. These engines were never re-numbered, No. 172 being scrapped in October, 1886, and No. 173 in June, 1884.

Towards the end of 1863, orders were placed with Beyer, Peacock & Co., of Gorton Foundry, Manchester, for the construction of a dozen coupled express engines. Numbered 178–189, *(Fig. 69)*. They were very fine engines, both for work and appearance, as might be expected from so renowned a firm of locomotive builders, much of the detail in external appearance being left entirely in their hands. The late George Beyer was always most particular in designing the curves of the framing, especially those of the leading end of the outside frames, chimney caps, dome casings, etc., the lines being curves and not radii. The first of these engines (Nos 178, 179 and 180) arrived in March, 1864, Nos 181–186 in April and Nos 187, 188 and 189

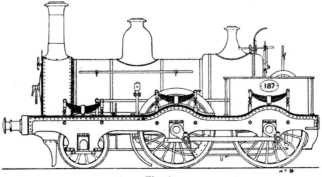

Fig. 69.

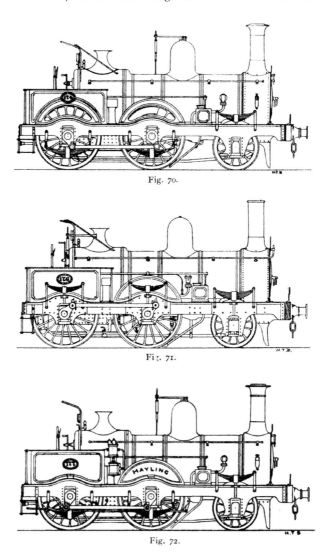

Fig. 70.

Fig. 71.

Fig. 72.

in May. The cylinders, which were placed horizontally, with centres 2 feet 6 inches apart, were 16 inch diameter by 20 inch stroke. The diameter of the blast pipe orifice was 4½ inches; the chimney stood 13 feet 2 inches above the rails. The slide valves were operated by the Stephenson link motion, with the weigh-bar shaft above. Both the inside and outside framings ran the whole length of the engine, and were of plate iron, the driving wheels having both inside and outside bearings. The leading wheels were 4 feet in diameter and the driving and trailing 6 feet. The wheelbase was 15 feet, being 7 feet 6 inches between each pair of centres. The boilers of these engines were 10 feet 6 inches long and 4 foot 1 inch outside diameter; the centre line being 6 feet 5 inches from the rails. They had double fireboxes, length outside 4 feet 10 inches, width 4 feet 1 inch, the inside box being secured to the crown of the shell by girder roof stay bars. The boiler contained 168 2-inch diameter tubes, and was provided with one injector and two pumps. They had double-beat regulators, a lock-up safety valve on the dome and two spring balance safety valves on the firebox. Stroudley removed the latter from all the engines, and with the exception of Nos 178 and 183, the spring balances were taken off the firebox and put on the dome. Their weight was 32 tons, of which 10 tons 4 cwt were on the leading wheels, 12 tons 16 cwt on the driving, and 9 tons on the trailing. These engines were all fitted by Stroudley with the Westinghouse air brake, No. 186 being the first coupled express engine of Craven's design to be so fitted. In 1889 Nos 185 and 189 were re-numbered 465 and 466, and were the only engines of this class whose numbers were altered. No. 181 was scrapped in December, 1886, No. 188 in March, 1887, Nos 182 and 187 in May, 1887, No. 186 in July, 1887, Nos 179 and 180 in December, 1887, Nos 183 and 184 in June, 1888, and No. 178 in December, 1889. This latter engine was running the Portsmouth train when Betchworth tunnel, from which it had just emerged, collapsed in the summer of 1887. Nos 465 and 466 were scrapped in 1890, or rather the wheels, motion, and framing were, for the boilers were used for supplying steam for the electric light plant which lights the Brighton Central Station and Lover's Walk goods yard and sidings. These have now been replaced by modern ones.

In June, 1864, four four-wheels coupled express engines were built at Brighton Works, numbered 174–177, *(Figs 70, 71 and 72)*. In their original state they had open splashers for the driving wheels, and open side sheets over the trailing wheels – around the latter and the splashers being double brass bands, as may be seen in *Fig. 70*. The chimneys had not originally copper tops, and were almost parallel, instead of having the rather large amount of taper usual on Craven's chimneys when they had no caps. One spring balance safety valve was on the dome, and two over the firebox. No. 174 was altered by Stroudley, having the trailing end of the framings lengthened, and the driving and trailing springs underhung, with the object of filling in the splashers and naming the engine, but this proposal was never carried into practice. The brass round the splashers, however, was retained, and although at one time it was painted over, the paint was afterwards removed and the bands polished. No. 174 was the only engine of this class to have the brass left on until it was broken up. On Saturday afternoon, 27 September, 1879, the firebox of this engine exploded at Lewes Station just as it was about to depart for London. The 2.05 p.m. train from Hastings arrived at Lewes a few minutes after three, and at this station the driver filled the tender tank with water. The train was booked to depart at 3.05 p.m., but it was not until ten minutes after that everything was ready, and the guard gave the signal to the driver to start. Then, just as the train began to move, there was a loud explosion, resembling the report of a gun, and for a few moments the engine and front part of the train was obscured from view by volumes of steam; but when this cleared off it was seen that the left-hand side of the firebox casing had been destroyed, the trailing end of the engine, the tender and guard's brake lifted bodily off the metals, and the smokebox door blown off, but the weather board had not been affected. The driver was found quite dead upon the roof of the carriage next to the guard's van. He had sustained dreadful injuries. His watch chain having blown across two or three pairs of rails into an adjoining field. The guard and fireman were badly injured, but not killed. At the time of the accident No. 174 had no safety valves on the dome, but two on the firebox. Nos 175 and

176 also had their framing lengthened, and the splashers and side sheets closed in, the brass strips around them being taken off, no doubt with the intention of naming the engines, but the springs remained over the axlebones as originally built, *(Fig. 71)*. Stroudley removed the balances from the domes of all of these engines, but again, at a later period, took them off the firebox and put two over the dome. No. 177 had the framing lengthened, and the bearing springs of the coupled wheels underhung, and at the same time was named *Hayling*. The framing of this class was of iron, and had thickening pieces round the horn blocks, and inside bearings of the driving wheels. The sandboxes, which were originally between the frames, were done away with, and new ones substituted, placed on the footplating just in front of the driving splashers. The cylinders were 16 inch diameter by 20 inch stroke; the diameter of the leading wheels was 4 feet, and of the driving and trailing wheels 6 feet. The wheelbase was 15 feet 3 inches, being 7 feet 9 inches from leading to trailing centres. The diameter of the boiler was 3 feet 11¾ inches, its length 10 feet 6 inches, and the height of its centre line above the rail level 6 feet 5 inches. The length of the outside firebox was 4 feet 9¾ inches. The heating surface of the tubes was 910 sq. ft, and of the firebox 94 sq. ft, making a total of 1,004 sq. ft. These engines weighed about 32 tons 13 cwt each, the weight being distributed as follows: On leading wheels 11 tons 8 cwt, on driving wheels 13 tons 1 cwt, and on trailing wheels 8 tons 4 cwt. The Westinghouse brake was fitted to all these locomotives. No. 174 was re-numbered 488 in July, 1881; was again re-numbered 454 in October, 1887, and in December, 1889, was scrapped. No. 175 was re-numbered 455 in 1890, and was scrapped in July, 1893. No. 176 was never re-numbered, and in October, 1890, was scrapped. No. 177, not re-numbered, was scrapped in June, 1889.

These engines were followed by two neat little singles, built at Brighton in June, 1864, for working between London and Tunbridge Wells, *via* Three Bridges and East Grinstead. We illustrate this class – numbered 190 and 191 – by *Fig. 73*, from which it will be noticed that they had outside bearings to all the wheels, the driving wheels being also provided with inside bearings, the inside frames finishing in front of the firebox casing. The chimneys of these engines were very conspicuous owing to their length. Brass bands went round the driving splashers. A few years after they were built Craven re-numbered them 24 and 33. The cylinders, which were placed horizontally,

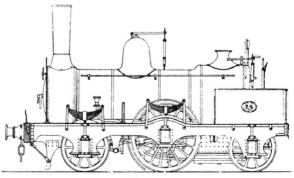

Fig. 73.

were 15 inch diameter, with a stroke of 20 inches, distance of centres apart 2 feet 7½ inches; the steam ports were 1⅜ inches wide by 10½ inches deep, and the exhaust ports 3½ inches wide. The leading and trailing wheels were 3 feet 9 inches in diameter, and the driving wheels 5 feet 6 inches. The wheelbase was only 13 feet 6 inches, being 6 feet 6 inches from the leading to driving centres, and 7 feet from the driving to trailing. From the centre of the leading wheels to the front of the buffer beam was 5 feet, and from the trailing centre to the end of the engine was 3 feet. The external diameter of the boiler barrel was 3 feet 11 inches, and its length 9 feet 9 inches, the centre line being 5 feet 11½ inches above the rail level. The steam dome was 1 foot 10 inches in diameter, and 2 feet 6 inches high. The length of the outside firebox was 4 feet 2 inches, the width 4 feet 3 inches, and the height inside 5 feet 1¼ inches. The regulator was of the double-beat type. There was a spring balance safety valve on the dome, and two over the firebox. The boiler was fed by two pumps worked by the crossheads. The slide valves were operated by the Allan straight link motion, having the weigh-bar shaft below. These engines were exceptionally heavy, their weight being about 38 tons 4 cwt, of which 12 tons 8 cwt was on the leading wheels, 13 tons 6 cwt on the driving wheels, and 12 tons 10 cwt on the trailing wheels. In October, 1875, No. 24 was re-numbered 267; in October, 1881, was again re-numbered 485, and scrapped in September, 1882. No. 33 was renumbered 280; in September, 1899, was again re-numbered 392, and in September, 1882, was scrapped.

At the early part of the year 1864, tenders were invited for the construction of twelve single-wheel express engines. The contract was given to Robert Stephenson & Co., of Newcastle-upon-Tyne, and they were delivered at Brighton between June and November, 1864, being numbered in the books of the contractors 1551–1562 and in the books of the railway company 194–205. A few months after all these engines had been running Stephenson & Co. re-purchased four of them – Nos 194, 196, 197 and 202 – owing to the fact that at that time they had a contract to supply the Egyptian Government Railways with some express engines of the same type. This railway was in a hurry for their delivery, and the builders not being able to supply the engines fast enough, fell back upon the Brighton Company. Owing to this, four other engines were built at Brighton at a subsequent date to replace them, and this affair has led to much complication in tracing the history of these engines built by Stephenson. It is generally supposed by most locomotive historians that four of these engines were converted into coupled engines by Stroudley, but as a matter of fact only two were so altered. In the meantime, we give the dimensions of these well-known locomotives, which are illustrated, as originally built, by *Fig. 74*: diameter of cylinders 16½ inches, stroke of pistons 22 inches, centres of cylinders apart 2 feet 6 inches; wheel-base 16 feet 4 inches, being 8 feet 4 inches from the leading to driving centres and 8 feet from the driving to trailing centres. The leading wheels were 4 feet 3 inches in diameter, driving wheels 6 feet 6 inches and trailing wheels 4 feet. The external diameter of the boiler barrel was 4 feet, length 11 feet 6 inches, centre from rail 6 feet 8½ inches. The steam domes were unusually large; in fact, they were the largest used on any of the Brighton engines. They were 2 feet 3 inches in diameter and 3 feet 4 inches high, the casing being bell-mouthed at the top. One spring balance safety valve was on the dome and two over the firebox.

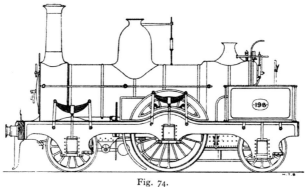

Fig. 74.

The front tube-plate was connected to the back-plate of the firebox casing by nine longitudinal stays. The boiler contained 152 brass tubes 2 inches in diameter and 11 feet 9 3/16 inches long; these were very long tubes, fitted into some of the longest barrels used on Brighton engines up to the present day, if we except the special 'long boiler' engines. The length of the outside firebox was 4 feet 8½ inches, the breadth 4 feet 0½ inch and the height inside 5 feet 4½ inches. The heating surface was as follows: tubes 936 sq. ft, firebox 87 sq. ft, total 1,023 sq. ft; firegrate area 13.75 sq. ft. The boiler was fed by one pump and one injector. The valve motion was of the Stephenson type, having the weigh-bar shaft below. These locomotives had both outside and inside framing, the latter extending the whole length of the engine; the driving wheels had inside as well as outside bearings. The weather boards were taper, the top being narrower than the bottom. This was the only instance on Brighton engines of weather boards tapering. The engines weighed about 31 tons 14 cwt each, the weight being distributed as follows: 11 tons 4 cwt on leading wheels, 12 tons 8 cwt on driving wheels and 8 tons 2 cwt on trailing wheels. No. 195 was named *Portsmouth* during Stroudley's time, and, like the other engines named, underwent the usual minor alterations. In the original state the reversing gear was on the right-hand side of the engine, but Stroudley put it on the left. The original driving wheels had twenty spokes, but the standard wheels, containing twenty-two, were adopted. The hand rails were placed higher up on the boiler and the safety valve taken off the dome, but at a later date both the safety valves were taken off the firebox and put on the dome. The injectors were discarded and replaced by pumps. In November, 1887, this engine was re-numbered 486, and in December, 1893, it was scrapped after being used for some time for pumping at Haywards Heath. No. 198 was converted by Stroudley into a four-wheels coupled express engine, and commenced to work as such in December, 1872, when it was re-numbered 205 and named *Kensington*. We illustrate this engine by *Fig. 75*. The new trailing ends of the frames were welded to the original portion that was retained, and the engine was fitted with Stroudley's standard type of boiler without a dome, and over the manhole were Adams' patent safety valves, resembling in appearance very much those introduced by Billinton on his latest express and tank engines. Subsequently these were removed and the boiler fitted with the standard dome and spring balances. The dimensions of this engine as rebuilt: Diameter of cylinders 17 inches, stroke of piston 24 inches, centre

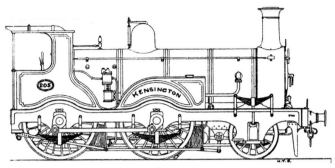

Fig. 75.

of leading to centre of driving wheels 8 feet 4 inches, from driving to trailing centres likewise 8 feet 4 inches, diameter of leading wheels 4 feet 3 inches and of driving and trailing 6 feet 6 inches. The boiler was telescopic and made in three rings. The mean diameter was 4 feet 1⅞ inches, length of barrel 10 feet 5 inches, centre line from rails 7 feet; tubes, 260, 1½ inch diameter and 10 feet 11¾ inches long; firebox casing, length 6 feet 2 3/16 inches, depth below boiler-centre line 4 feet 11 inches. This class of boiler was the only one of Stroudley's design, excepting those of the 'Terriers' and the 'Sussex', having the depth of the firebox the same at the front as at the back. The heating surface of the tubes was 1,112 sq. ft and of the firebox 112 sq. ft, making a total of 1,224 sq. ft. The weight of this engine in working trim was about 39 tons 5 cwt, the leading wheels carrying 12 tons 13 cwt, the driving wheels 15 tons 1 cwt and trailing wheels 11 tons 11 cwt. In March, 1897, No. 205 was re-numbered 505, and in January, 1898, was re-numbered 605, and in May, 1901, was scrapped. No. 199 was named *Paris*, and was the first locomotive to be named under Stroudley, but at a later date, when the new *Paris*, No. 313, was built, this engine was re-named *Newport*. In November, 1887, it was re-numbered 489 and was scrapped in 1893. No. 200 was named *Dieppe, (Fig. 76)*, which shows this class as modified. This engine was the only one that did not have the safety valves removed from the firebox on to the dome. It was re-numbered 490 in November, 1887, and scrapped in April, 1896. Previous to being scrapped it was used for pumping purposes for a long time at the locomotive shed at Tunbridge Wells.

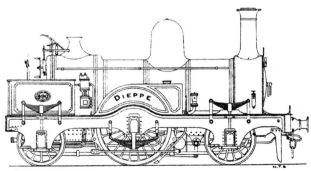

Fig. 76.

No. 201 was re-numbered 111, and again later on 197, when it was named *Cavendish*, but when Stroudley's No. 211, *Beaconsfield* was re-named *Cavendish*, No. 197 was stopped from working and stood in the old coppersmith's shop at the works at Brighton till 1887, when it was done up, and commenced work again as No. 487, *Chichester*, and in 1893 it was sold. In June, 1871, No. 203 was rebuilt by Stroudley and named *Sussex*; it had a new boiler and cylinders, the framing at the trailing end was lengthened and a cab was put over the footplate *(Fig. 77)*. The cylinders were 17 inch diameter, stroke of pistons 23 inches, a very curious dimension for an English locomotive, this being the only instance upon the line. The centres of the cylinders were 2 feet 6 inches apart. The frames at the trailing end were lengthened from 3 feet 1 inch to 4 feet 4 inches, the trailing axle being under the firebox. The external diameter of the boiler was 4 feet 3 inches, the length was 11 feet, and the centre line from the rail level was 7 feet. This engine had no dome, but over the manhole, which was on the third ring of the barrel, were mounted Adams' safety valves, the same as on No. 205. These safety valves were discarded a few years after, not having proved a success, and the usual dome and spring balances substituted. In the manhole was the regulator, of a type having a circular face, with four openings. This regulator was copied from one introduced in the early days by Sharp, Roberts & Co., and it is still used as the standard pattern. The boiler contained 262 tubes 1½ inches in diameter and 11 feet 6 inches long. The length of the outside firebox was 6 feet 2 3/16 inches, the breadth 4 feet 0½ inch and depth at front and back below the centre line of the boiler 4 feet 5 inches. The heating surface of the tubes was 1,183 sq. ft and of the firebox 105 sq. ft, giving a total of 1,288 sq. ft. The weight of this engine was 36 tons 6 cwt in working order, having 11 tons 12 cwt on the leading wheels, 14 tons 14 cwt on the driving wheels and 10 tons on the trailing wheels. The inside bearing springs, which originally were of plates, were replaced by two volute springs to each bearing. The *Sussex* has long been known as a record engine for speed: In February, 1877, this engine left Brighton for London 10 minutes behind time, having fifteen coaches on for London Bridge and three for Victoria, and arrived in London one minute before time, having made up eleven minutes in 50½ miles, with eighteen coaches on for a distance of 40 miles.

In 1892 this engine was fitted with a new valve motion, which was devised by the late David Joy, already well-known as the inventor of the 'Joy' valve gear. In this novel

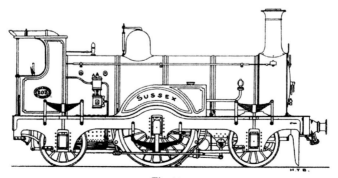

Fig. 77.

valve motion a single eccentric is used for each cylinder, and is shifted transversely across the axle from side to side for forward or backward gear. For this purpose the axle is squared where the eccentric is fixed upon it, and an oblong slot through the eccentric allows it to be slid across from side to side of the axle. At each end of this slot is formed a small cylinder, which works over a corresponding ram fixed on opposite sides of the square of the axle. It only remains to force oil or other fluid under pressure into either end of the axle, and thence into one or other of the two small cylinders, according as it is required to move the eccentric into either position for forward or backward gear, or to hold it between these extremes for any point of expansion or for mid-gear. The oil is made to pass into either end of the axle by a small cylinder, placed on the footplate of the engine, and fitted with a piston which is moved either by a hand wheel and screw or by steam.

When this engine was rebuilt the original boiler was put into No. 119 engine, one of Craven's goods engines. In January, 1897, No. 203 was re-numbered 503, and was scrapped in January, 1899. In July, 1872, No. 204 was converted into a coupled engine, like No. 205, and named *Westminster*. In February, 1897, it was re-numbered 504, and scrapped January, 1899. The original No. 205 was re-numbered 198, when the original 198 was converted into a coupled engine and re-numbered 205, this alteration being made in order to keep the class together. This new 198 was named *Drayton*, and underwent the same alterations as the others of its class, which were not rebuilt; in November, 1887, it was re-numbered 488, and was scrapped in June, 1894. All these engines were fitted with the Westinghouse air brake at some time by Stroudley.

As the history of this class owing to the re-numbering and conversions is somewhat complicated, this statement may help more fully to make the matter clear.

194	Purchased by Egyptian Government Railways.
195	Named *Portsmouth*, re-numbered 486.
196	Purchased by Egyptian Government Railways.
197	Purchased by Egyptian Government Railways.
198	Converted into four-coupled engine 1872, re-numbered 205 *Kensington*, re-numbered 505.
199	Named Paris, re-named *Newport*, re-numbered 489.
200	Named *Dieppe*, re-numbered 490.
201	Re-numbered 111, re-numbered 197, named *Cavendish*, re-numbered 487 *Chichester*.
202	Purchased by Egyptian Government Railways.
203	Rebuilt 1871 and named *Sussex*, re-numbered 503.
204	Converted into four-coupled engine, 1872, named *Westminster*, re-numbered 504.
205	Re-numbered 198, named *Drayton*, re-numbered 488.

In July, 1864, two six-wheels coupled goods engines were built at Brighton, numbered 192 and 193, *(Fig. 78)*. The cylinders were 16 inch diameter with a stroke of 24 inches, their centres being 2 feet 6 inches apart. The steam ports were 1 7/16 inches wide and 14 inches long, the exhaust ports being 3 inches by 14 inches. The boiler was made in three rings, the largest external diameter being 4 feet 2 inches, and the length of the barrel 11 feet. The centre line was 6 feet 6½ inches above the rail level, and the steam dome, which was placed on the middle ring, was 2 feet 2 inches in diameter outside and 3 feet 3 inches high. The height of the chimney from the rails was 13 feet 2 inches.

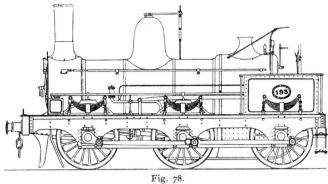

Fig. 78.

The boilers each contained 160 brass tubes 2 inches in diameter and 11 feet 5 inches in length. The length of the outside firebox was 4 feet 10 inches, the height inside 4 feet 10½ inches, and the breadth outside 4 feet 1 inch. The heating surface of the tubes was 944 sq. ft, and of the firebox 92 sq., ft, making a total of 1,036 sq. ft. The wheels were 5 foot 1 inch diameter on tread. From the front of the buffer beam to the centre of the leading wheels was 5 feet 4½ inches, from the leading to the driving centres 7 feet 9 inches, from the driving to the trailing centres 8 feet – the total wheelbase being 15 feet 9 inches. From the trailing centre to the end of the engine was 3 feet 7 inches. The slide valves were worked by Allan's straight link motion, having the weigh-bar shaft above. The boiler was fed by two long-stroke pumps worked by the engine crossheads. Steam was admitted to the cylinders by a double-beat regulator placed in the dome. The inside and outside framings were of iron, having outside bearings to all the wheels, with inside ones also to the drivers. The inside frames extended the whole length of the engine. One spring balance safety valve was placed on the dome and two over the firebox. These engines averaged in weight about 33 tons 17 cwt each, having 11 tons 13 cwt on the leading wheels, 13 tons 10 cwt on the driving wheels, and 8 tons 14 cwt on the trailing wheels. No. 192 was re-numbered 382 and afterwards 475, and scrapped in August, 1897. No. 193 was re-numbered 383, and later on 476, and scrapped in December, 1894.

These two engines were followed by another goods engine, built at Brighton in September, 1864, and was numbered 54, *(Fig. 79)*. Although much smaller than Nos 192 and 193, it had the same diameter and stroke of cylinders and amount of heating surface. It was a neat little engine, and did some wonderfully good work in its time. The cylinders were placed at an inclination of 1 3/16 in 12, their centres being 2 feet 6 inches apart. The diameter of the blast-pipe orifice was 4½ inches. From the front of the buffer plank to the centre of the leading wheels was 6 feet 6½ inches, from the leading to the driving centres was 7 feet 3½ inches, from the driving to the trailing centres 8 feet, making a total wheelbase of 15 feet 3½ inches. From the trailing centres to the end of the framings was 2 feet 9 inches. The diameter of the wheels on tread was 4 feet 9 inches. The largest external diameter of the boiler barrel was 3 feet 9 inches, and its length 11 feet 6 inches. The height of the boiler centre line above the rail level was 6 feet 3 inches, and of the chimney 13 feet. The boiler contained 151 brass tubes

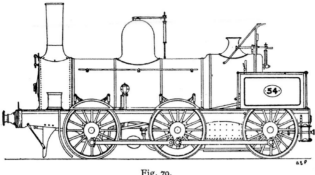

Fig. 79.

2 inches in diameter and 11 feet 11 inches long. The steam dome was very large, being 2 feet 2 inches in diameter and 3 feet 4 inches in height. The length of the outside firebox was 4 feet 10¾ inches, the height inside 5 feet 3½ inches, and the breadth outside 4 feet 0½ inch. The tubes contained 945 sq. ft of heating surface and the firebox 91 sq. ft, giving a total of 1,036 sq. ft. The boiler was fed by two long-stroke pumps, the pump rams being connected to the crossheads. The steam regulator was of the double-beat pattern. The framings were of iron, and by referring to the illustration it will be seen that they were inside, or placed between the wheels, this type of engine being of very rare occurrence during Craven's period. The motion was of the ordinary Stephenson link type with the weigh-bar shaft below. The weight was distributed as follows: 11 tons 4 cwt on the leading wheels, 12 tons 4 cwt on the driving wheels, and 7 tons 3 cwt on the trailing wheels, giving a total of 30 tons 11 cwt. This engine was re-numbered by Stroudley 136, and in November, 1878, was again re-numbered 381. In November, 1885, it was purchased by a firm of contractors.

No. 54 was followed by half-a-dozen more goods engines, numbered 206–211, *Fig. 80* represents this class as fitted with injectors by Stroudley. These locomotives were built in the following order: Nos 206 and 207 in December, 1864, Nos 208 and 209 in February, 1865, and Nos 210 and 211 in March, 1865. The cylinders were 16 inch diameter and 24 inches in stroke, they were placed at an inclination of 1 3/16 in 12, with the centres 2 feet 7 inches apart. The steam ports were 1⅜ inches wide and

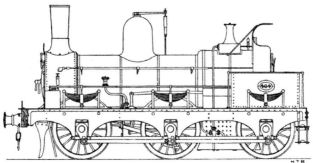

Fig. 80.

14 inches long, the exhaust ports being 3 1/16 inches by 14 inches. The six-coupled driving wheels were 5 feet 0 inches in diameter. From the front of the buffer beam to the centre of the leading wheels was 5 feet 3 inches, from the leading to the driving centres 7 feet 8 inches, from the driving to the trailing centres 8 feet 3 inches, making a total wheelbase of 15 feet 11 inches. From the trailing centres to the end of the engine was 3 feet 7 inches. The external diameter of the boilers was 4 feet 2 inches, the length 10 feet 6 inches, the centre line being 6 feet 7 inches above rail level. They contained 154 brass tubes 10 feet 11½ inches long and 2 inches in diameter. The length of the firebox casing was 5 feet 4 inches, the height inside 5 feet 6½ inches, and the breadth outside 4 feet 1 inch. The heating surface of the tubes was 870 sq. ft, and that of the firebox 106 sq. ft, giving a total of 976 sq. ft. The steam dome was 2 feet 2½ inches in diameter and 3 feet 1½ inches high. The front tube plate was connected to the back plate of the firebox casing by nine longitudinal stays. One spring balance safety valve was on the dome, and two over the firebox. Stroudley afterwards altered the position of these safety valves similarly to those on the other engines. These locomotives had double-beat regulators of the kind introduced by the late Ramsbottom on the LNWR. The boilers were supplied with water by two pumps. A year or so previous to Stroudley's decease Nos 206, 209 and 210 – but not during the time they had these numbers – were fitted with injectors, the original pumps being removed. The frames, both inside and outside, were of iron, and extended the whole length of the engine. The driving wheels had inside and outside axleboxes, the other wheels having outside only. The slide valves were actuated by Allan's straight link motion, having the weigh-bar shaft below. These engines weighed about 34 tons 4 cwt each, having the weight distributed as follows – 11 tons 14 cwt on the leading wheels, 13 tons 8 cwt on the driving wheels, and 9 tons 2 cwt on the trailing wheels. In 1875 No. 206 was re-numbered 271, in May, 1880, it was again re-numbered 396, and in March, 1896, became 476; in September, 1896, it was scrapped. No. 207 was renumbered 272 in 1875, in April, 1880, it was altered to 397, and in April, 1895, was scrapped. No. 208 was re-numbered 380 in November, 1878, and again later on 474, being scrapped in April, 1896. No. 209 was re-numbered 387 in June, 1879, and again afterwards 463, and broken up in August, 1897. No. 210 was re-numbered 388 in June, 1879, and again altered to 464. This engine and No. 387 were the last goods engines to have the safety valves removed from the crown of the firebox and put on the dome, this being done when they were fitted with the injectors. No. 464 was also broken up in August, 1897. No. 211 was renumbered 394 in February, 1880, and scrapped in April, 1895.

The next class of locomotive to be constructed at Brighton was a front-coupled side-tank engine, built in July, 1865, and numbered 212, *(Fig. 81)*. It was constructed for working between South Croydon and Victoria, and was stationed at Battersea yard. These are its principal dimensions: Diameter of cylinders 16 inches, stroke of pistons 20 inches, centres of cylinders 2 feet 5 inches apart, inclination of cylinders 1⅜ in 13½; the steam ports were 1½ inches wide and 13 inches long, the exhaust ports being 3½ inches by 13 inches. From the front of the leading buffer beam to the centre of leading wheels was 5 feet 8 inches, from the leading centres to the driving 6 feet 3 inches, and from the driving to the trailing centres 8 feet 1 inch, making a total wheel-base of 14

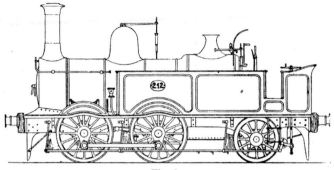

Fig. 81.

feet 4 inches, the overhang at the trailing end being 5 feet 2 inches. The diameter of the leading and driving wheels was 5 feet, and of the trailing wheels 3 feet 9 inches. The boiler was made in three rings, and was 3 feet 11 inches in diameter and 9 feet 9 inches in length. Its centre line was 6 feet 3½ inches, and the top of the chimney was 13 feet 2 inches above the rails. The steam dome was 1 foot 9 inches in diameter and 2 feet 7 inches high. The boiler contained 158 tubes 2 inches in diameter and 10 feet 2 inches long. The length of the firebox casing was 4 feet 8 inches, the height inside 5 feet 1 inch, and the width outside 4 feet 1 inch. The heating surface of the tubes was 829 sq. ft, and of the firebox 86 sq. ft, giving a total of 915 sq. ft. The boiler was fed by two long-stroke pumps, and had a double-beat regulator. The safety valves were in the usual position, one on the dome and two over the firebox. The side tanks carried 355 gallons of water, and the tank under the footplate, behind the trailing axle, 290 gallons. The engine not being able to carry sufficient water with the above arrangement of tanks, another one was added in July, 1868, fitted up in front and extending under the cylinders. This tank carried 155 gallons, thus increasing the total water capacity to 800 gallons. The bunker would carry 1 ton 5 cwt of coke. The weight of the engine was 33 tons 10 cwt, the leading wheels carrying 11 tons 6 cwt, the driving 12 tons 10 cwt, and the trailing wheels 9 tons 14 cwt. The framing was of iron, and inside, as seen from the illustration. Renumbered 413 in March, 1880, it was scrapped in the following May.

In July, 1865, Craven built at Brighton a four-wheels coupled side tank engine for working on the Midhurst branch, but owing to its weight, especially on the trailing end, the Government inspector would not allow it to run on that line, and in consequence it was put on the Eastbourne branch, where the rails were heavier. This loco was numbered 213, *(Fig. 82)*. It will be seen that it had outside framings, and the leading and driving wheels were coupled. The engine practically represented Craven's standard design of tank engine, for although there were not many of the class, others were built at a later date which were very much the same. The cylinders, which were placed at an inclination of 1 3/16 in 12, and 2 feet 5 inches between centres, were 15 inches in diameter by 20 inch stroke, the steam ports being 1⅛ inches wide by 14 inches long, and the exhaust ports 3 inches by 14 inches. The coupled wheels were 5 feet in diameter on tread and the trailing wheels 3 feet 9 inches. From the front buffer beam to the leading wheel centre was 4 feet 8 inches, from leading to driving centres was 7 feet 3 inches, from driving to

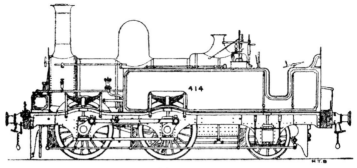

Fig. 82.

trailing centres 8 feet 7 inches, and the overhang at the trailing end was 4 feet 7 inches. The boiler, which was of the usual design with three rings, having the largest ring in the middle, was 3 feet 11 inches in external diameter and 9 feet 9 inches long. The boiler centre line was 6 feet 5 inches, and the height of the chimney 13 feet, above the rails. The steam dome was 2 feet in diameter outside and 3 feet 2½ inches high, and contained a double-beat regulator. The front tube plate was connected to the back plate of the firebox by nine longitudinal stays. The boiler contained 158 tubes 2 inches in diameter and 10 feet 2 inches long. The length of the outside firebox was 4 feet 8 inches, the breadth 4 feet 1 inch, and the depth inside 5 feet 2 inches. The tubes contained 829 sq. ft of heating surface, and the firebox 86 sq. ft, giving a total of 915 sq. ft. The coal bunker carried 1 ton 5 cwt, the side tanks combined held 545 gallons of water, and the tank under the footplate behind the trailing axle 257 gallons, making a total of 802 gallons. The boiler was supplied with water by one pump worked by the crosshead, and one injector. The valve motion was of the usual Stephenson type, having the weigh-bar shaft above, but in front of, the links.

This engine was fitted with a conical wire spark arrester fixed on the top of the blast pipe and extending to the base of the chimney. The inside frames were of iron, running the whole length of the engine. The outside ones had thickening pieces round the horn blocks, the driving wheels having both inside and outside bearings. The hand brake acted on all the wheels. The sandboxes were originally behind the side tanks, but Stroudley removed them to the front of the smoke box. He also took the safety valve off the dome, leaving the other two over the firebox. The trailing buffer beam was subsequently duplicated to enable a tool-box to be placed behind the bunker. No. 213 weighed 36 tons 8 cwt, distributed as follows: 11 tons 6 cwt, on the leading wheels, 13 tons 6 cwt on the driving wheels and 11 tons 16 cwt on the trailers. In March, 1880, this engine was re-numbered 414, and in September, 1882, it was scrapped.

In September, 1865, two small single engines were built at Brighton, numbered 29 and 30. No. 29 was stationed at Battersea, but No. 30 stood at Horsham for working the Guildford branch. They are illustrated by *Fig. 83*. The cylinders, which were placed horizontally and 2 feet 5 inches between centres, were 15 inches in diameter by 20 inch stroke. The steam ports were 1⅛ inches wide and 14 inches long, the exhaust ports being 3 inches by 14 inches. The leading and trailing wheels were 3 feet 9

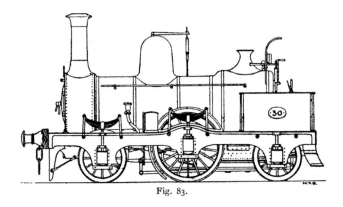

Fig. 83.

inches in diameter and the driving wheels 5 feet 6 inches. The wheelbase was 13 feet 10 inches, being 6 feet 10 inches from leading to driving centres and 7 feet from driving to trailing. The external diameter of the boiler barrel was 3 feet 11 inches, the length 9 feet 9 inches; it contained 162 tubes 2 inches in diameter. The centre line of the boiler was 6 feet 1¼ inches from the rails. The length of the outside firebox was 4 feet 4 inches, width 4 feet 1 inch and height 5 feet 1½ inches. The tubes contained 812 sq. ft of heating surface and the firebox 75 sq. ft, giving a total of 887 sq. ft. The boiler was fed by one injector and one pump. Steam was admitted to the cylinders by a double-beat regulator placed in the dome. One spring balance safety valve was on the dome and two more over the firebox. The slide valves were worked by the Stephenson link motion, having the weigh-bar shaft beneath. As will be seen from the illustration, these engines had outside frames, which, as well as the inside ones, were of iron, and extended the whole length of the engine. The driving wheels had both inside and outside bearings. The outside framing had thickening pieces round the axle-box horns. The average weight of the engines was about 27 tons 14 cwt each, distributed as follows: 9 tons on the leading wheels, 12 tons 12 cwt on the driving wheels, and 6 tons 2 cwt on the trailers. No. 29 was eventually re-numbered 293; in October, 1877, it was again re-numbered 366; and was taken from service in September, 1885. No. 30 was re-numbered 19; in December, 1875, it was re-numbered 263; in July, 1881, it was altered to 486; and finally withdrawn in May, 1886.

In November, 1865, two large single driving wheel engines were built at Brighton to replace two of the Stephenson singles sent to Egypt. They were numbered 194 and 196, *(Fig. 84)*. The cylinders, which were horizontal and 2 feet 6 inches between the centres, had a diameter of 17 inches and a stroke of 22 inches. The steam ports were 1½ inches wide by 17 inches long, the exhaust ports being 4 inches by 17 inches. The leading wheels were 4 feet 3 inches in diameter, the driving 6 feet 6 inches, and the trailing wheels 4 feet. The wheel-base was 16 feet 3 inches, being 8 feet 2-in. from leading to driving centres, and 8 feet 1 inch from driving to trailing centres. The boiler was of the usual design, the barrel being 4 feet in diameter outside the middle ring and 11 feet 6 inches long; it contained 192 tubes, 2 inches in diameter and 11 feet 11 inches in length; nine longitudinal stays connected the tube and back plates. The steam dome was very large, being 2 feet 2 inches in diameter and 3 feet 5 inches high. The chimney

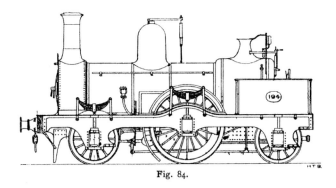

Fig. 84.

top was 13 feet, and the centre line of the boiler 6 feet 6 inches above the rail level. The length of the outside firebox was 4 feet 11½ inches, and the width 4 feet 3 inches, the height inside being 5 feet 7 inches. The tubes contained 870 sq. ft of heating surface, and the firebox 95 sq. ft, giving a total of 965 sq. ft. The weight of these engines was about 30 tons each, the leading wheels carrying 10 tons 13 cwt, the driving 12 tons 16 cwt, and the trailing 6 tons 11 cwt. A double-beat regulator admitted steam to the cylinders, but this was afterwards replaced by Stroudley's standard pattern, which was fitted to the majority of Craven's engines. The valve gear was of the Stephenson type, with the weigh-bar shaft below the motion. One injector and one long-stroke pump supplied the boiler with water. One spring balance safety valve was on the dome and two over the firebox, but these were afterwards altered as in the other engines previously mentioned. These locos were subsequently named *Glynde* and *Pevensey* respectively, and underwent the usual Stroudley alterations, including removal of the reversing rod from the right hand to the left and the replacement of the injector by a pump. They were also fitted with the Westinghouse air brake. The outside framings were 1⅛ inches thick; the inside framings only went as far as the front of the firebox casing. These two engines were exactly alike with the exception of the weather boards, that of No. 194 having the top turned over and that of No. 196 being quite vertical. They were later re-numbered 484 and 485, and scrapped in 1892 and 1890 respectively.

In November, 1865, two four-wheels coupled tank engines were built at Brighton

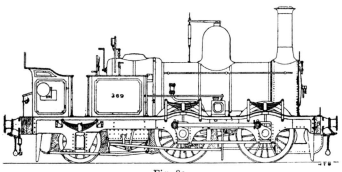

Fig. 85.

for working on the South London line. Numbered 214 and 215, and illustrated by *Fig. 85*, from which it will be seen that they had no side tanks, but a small tank under the cylinders, another under the coke box, and a third between the trailing wheels underneath the footplate. The cylinders were 15 inch diameter, with a stroke of 20 inches, and were placed at an inclination of 1¼ in 12, with their centres 2 feet 5 inches apart. The steam ports were 1⅛ inches wide and 14 inches deep, the exhaust ports being 3 inches by 14 inches. The leading and driving wheels were 5 feet in diameter, and the trailing 3 feet 9 inches. The wheelbase was 15 feet 10 inches, between leading and driving centres being 7 feet 3 inches, and between driving and trailing 8 feet 7 inches. The overhang at the leading end was 4 feet 10 inches, and at the trailing end 4 feet 9 inches. The external diameter of the boiler was 3 feet 11 inches, its length being 9 feet 9 inches, and it contained 158 tubes of 2 inch diameter and 10 feet 2 inches long; its centre line was 6 feet 5½ inches above the rails. The dome was 2 feet in diameter outside, and 3 feet 3 inches high. The heating surface of the tubes was 829 sq. ft, and that of the firebox 86 sq. ft, giving a total of 915 sq. ft. The length of the outside firebox was 4 feet 8 inches, and the width 4 feet 1 inch, the height inside being 5 feet 2½ inches. The boiler was fed by one pump and one injector. The motion was of the Stephenson type, having the weigh-bar shaft above. The inside frames were of iron, and ran the whole length of the engine; the outside frame plates had thickening pieces round the horn blocks. Two spring balances were on the dome, but unlike the usual Craven practice there were no safety valves over the firebox, the manhole being covered by a flat casing similar to that on Billinton's earlier engines. Another peculiar feature of these engines was that the trailing springs were on the top of the axlebox, but below the footplating; the position of the clack box on the boiler was also unusual. The hand brake was fitted to the driving and trailing wheels, having two blocks to each wheel. In May, 1878, No. 214 was re-numbered 369, and in September, 1882, it was scrapped. No. 215 was first renumbered 371, then 497, and in June, 1884, was broken up. When new, both of these engines were stationed at New Cross sheds.

In March, 1866, Craven purchased from Manning, Wardle & Co., of the Boyne Engine Works, Leeds, two ballast engines for conveying material for several new lines then being made for the LBSCR system. These engines were numbered 219 and 220, and when new were painted a dark green. They proved very good engines, and were afterwards used on the main line goods, No. 220 for a long time working the fast 5.05 p.m. Willow Walk and Lewes goods train. The cylinders, which were 16 inch diameter, with a stroke of 22 inches, were placed 2 feet 5 inches apart between centres. The wheels were 4 feet 6 inches in diameter, and the wheel-base 14 feet 3 inches, being 7 feet 1 inch from the leading to the driving centres, and 7 feet 2 inches from the driving to the trailing. The leading overhang was 5 feet 2 inches, and the trailing 3 feet 1½ inches. The boiler was made in three rings, and had a diameter of 3 feet 9 inches outside, the length being 10 feet and it was fed by two injectors. The length of the outside firebox was 4 feet 7 inches, its breadth 3 feet 7½ inches, and its depth below the centre line of the barrel 4 feet 8½ inches. Referring to *Fig. 86*, it will be seen that these engines had a fluted dome, mounted on a square seating, and also a fluted safety valve funnel. No. 219 had a copper top to this funnel, and No. 220 a brass one. The frames were inside,

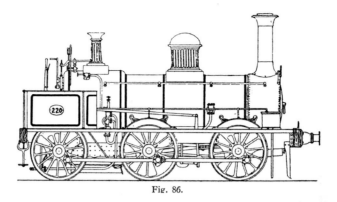

Fig. 86.

of iron plate, the springs being above the axle boxes. These engines had the Stephenson valve motion, with the weigh-bar shaft below, and their weight was 35 tons 9 cwt, distributed as follows: 11 tons 8 cwt on the leading wheels, 12 tons 3 cwt on the driving, and 11 tons 18 cwt on the trailing wheels. No. 219 was re-numbered 386, and in October, 1885, was sold to the West Lancashire Railway, by whom it was named *Brighton*, and numbered 5. About three years ago it was seen at Old Trafford Park sidings, Manchester, with a MSSLR type of chimney. No. 220 was never re-numbered. In November, 1887, it was scrapped. Stroudley replaced the fluted dome and safety valve column of this engine by a plain casing similar to those on Craven's engines.

In April, 1866, four side tank engines were finished at the Brighton Works for working the South London line; numbered 17, 216, 217 and 218, *(Fig. 87)*. The cylinders were 15 inches in diameter, and the stroke of pistons was 20 inches; they were placed 2 feet 5 inches apart, and at an angle of 1¼ in 12. The steam ports were 1⅛ inches wide and 14 inches long, the exhaust ports being 3 inches by 14 inches. The coupled wheels were 5 feet in diameter, and the trailing 3 feet 9 inches. The wheelbase was 15 feet 10 inches, being 7 feet 3 inches from the leading to the driving centres, and 8 feet 7 inches from the driving to the trailing. The leading overhang was 4 feet 10 inches, and the trailing 4 feet 9 inches. The diameter of the boiler barrel was 3 feet 11 inches, and the length 9 feet 9 inches, the centre line being 6 feet 5½ inches from the rails. The boiler contained 158 tubes, 2 inches in diameter by 10 feet 2 inches in length.

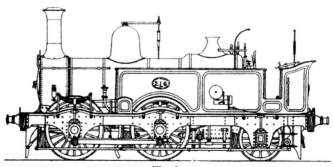

Fig. 87.

The steam dome was 2 feet in diameter and 3 feet 3 inches high. The length of the outside firebox was 4 feet 8 inches, the width 4 feet 1 inch, and the height inside 5 feet 2½ inches. The heating surface of tubes was 829 sq. ft, and of the firebox 86, giving a total of 915 sq. ft. The boiler was fed by one pump and one injector. The side tanks carried 630 gallons of water, and the tank under the trailing end of the footplate 270 gallons, giving a total of 900 gallons. The bunker carried 2 tons 5 cwt of coke. These engines had both inside and outside framing, and valve gear of the Stephenson type, having the weigh-bar shaft above. The weight was 38 tons 2 cwt, of which 11 tons 14 cwt were on the leading wheels, 13 tons 10 cwt on the driving wheels, and 12 tons 18 cwt on the trailing wheels. When new, Nos 17 and 217 were sent to Battersea, and Nos 216 and 218 to New Cross. In April, 1875, No. 17 was re-numbered 261, and in October, 1881, was again re-numbered 465; in November, 1886, it was broken up. No. 216 was re-numbered first 376, and afterwards 498. No. 217 was re-numbered 377, then 499, and in January, 1885, was scrapped.

These engines were followed in April, 1866, by a large goods engine, No. 221, (*Fig. 88*), built for working goods trains between Newhaven Wharf and Willow Walk. The cylinders, which were 16 inches in diameter, with a stroke of 24 inches, were placed 2 feet 7 inches apart; their diameter, however, was afterwards increased to 17 inches. The diameter of the blast pipe orifice was 5 inches. The wheels were 5 feet in diameter, and the wheel-base was 15 feet 11 inches, being 7 feet 8 inches from the leading to the driving centres, and 8 feet 3 inches from the driving to the trailing. The leading overhang was 5 feet 2 inches, and the trailing 4 feet 9 inches. The boiler was made in three rings, the largest external diameter being 4 feet 2 inches, and the length 10 feet 6 inches, whilst the centre line was 6 feet 7 inches above the rail level. The boiler contained 208 2 inch tubes. This engine had a double firebox, similar to those then used by J. Cudworth on the SER, having a longitudinal mid-feather; its length was 8 feet, width 4 feet 1 inch, and height inside at the front end 5 feet 3½ inches, and at the back 2 feet 3½ inches. The heating surface of the tubes was 1,060 sq. ft, and of the firebox 148 sq. ft, giving a total of 1,208 sq. ft. The front tubeplate and the firebox back-plate were connected by eight longitudinal stays. The crown of the inside firebox was connected to the outer casing by sixteen girder roof stay bars placed athwart the box. Steam was admitted to the cylinders by a double-beat regulator placed in the

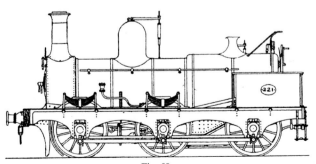

Fig. 88.

dome. One spring balance safety valve was on the dome, and two were over the crown of the firebox. The valves were operated by the Stephenson link motion, having the weigh-bar shaft below. The inside frames, which were of iron plate, extended the full length of the engine. The weight was distributed as follows: 12 tons 10 cwt on the leading wheels, 13 tons on the driving, and 12 tons on the trailing, giving a total of 37 tons 10 cwt. This engine was afterwards re-numbered 389, and in April, 1886, it was broken up.

In May, 1866, Craven built two singlewheel side tank engines for working between Croydon and Wimbledon and the Crystal Palace. Numbered 222 and 223, *(Fig. 89)*. The cylinders, which were placed horizontally and 2 feet 6 inches between centres, were 15 inches in diameter with a 20 inch stroke of the pistons. The steam ports were 1½ inches by 14 inches, and the exhaust ports 3 inches by 14 inches. The diameter of the leading and trailing wheels was 3 feet 6 inches, and of the drivers 5 feet. The leading and trailing wheels had outside, and the drivers inside bearings. The wheelbase was 14 feet 6 inches, equally divided. From the front of the buffer beam to the leading wheel centre was 4 feet 9 inches, the trailing overhang being 4 feet 10 inches. The boiler was 3 feet 7½ inches in diameter and 9 feet 10 inches long, with its centre line only 5 feet 9 inches above the rails. Length of the outside firebox was 4 feet, the breadth 4 feet 1 inch, and the height inside 4 feet 6 inches. The front tubeplate was connected to the firebox back plate by eight longitudinal stays. Outside diameter of the dome was 2 feet and the height 2 feet 9 inches. The boiler contained 147 tubes of 2 inch diameter, 10 feet 3 inches long, giving a heating surface of 750 sq. ft, whilst that of the firebox was 72 sq. ft, making a total of 822 sq. ft. These engines were fitted with double-beat regulators and the Stephenson link motion, the weigh-bar shaft being below. Side tanks carried 600 gallons, and the tank under the bunker 280 gallons. The weight was distributed as: 10 tons 1 cwt on the leading wheels, 11 tons 5 cwt on the driving, and 9 tons 17 cwt on the trailing wheels, making a total of 31 tons 3 cwt. No. 222 was rebuilt by Stroudley, who put in a new standard boiler and added a cab, the trailing end of the frames being lengthened. They were never re-numbered, but No. 222 was named *Egmont*. It was broken up in January, 1884, and No. 223 in March, 1882.

These were followed in June, 1866, by two small single-wheel engines, numbered 197 and 202, and built to replace Nos 197 and 202 of Stephenson's build that went

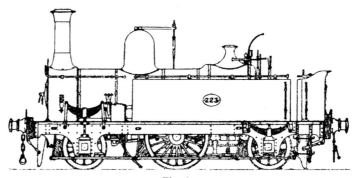

Fig. 89.

out to Egypt. The cylinders were 15 inches in diameter and the stroke of the pistons 20 inches; they were placed horizontally, with the centres 2 feet 5 inches apart. The steam ports were 1⅛ inches by 14 inches, and the exhaust ports 3 inches by 14 inches. The leading and trailing wheels were 4 feet in diameter, and the driving 6 feet. The wheelbase was 14 feet 6 inches, equally divided. The leading overhang was 4 feet 9 inches and the trailing 3 feet 1 inch. The boiler was 3 feet 11 inches in diameter and 9 feet 9 inches long, the centre line being 6 feet 4¼ inches above rail level. The steam dome was 2 feet in diameter and 2 feet 11 inches high. The length of the outside firebox was 4 feet 8 inches, the breadth 4 feet 2 inches, and the height inside 5 feet 1½ inches. The heating surface of the tubes was 840 sq. ft, and that of the firebox 90 sq. ft, making a total of 930 sq. ft. The valves were worked by the Stephenson link motion, having the weigh-bar shaft below. These engines had inside and outside framings, the driving axle having bearings in each. Thickening pieces of iron plate were added round the axle box horns. The weight of the engines was about 29 tons 12 cwt, distributed as follows: on the leading wheels 9 tons 14 cwt, on the driving 12 tons 18 cwt, and on the trailing wheels 7 tons. No. 197 was named *Solent* by Stroudley, and was re-numbered 266 in December, 1876, and 484 in October, 1881. In April, 1886, it was broken up. No. 202 was not named, but was re-numbered 112 in 1872, and 354 in February, 1877. In July, 1884, it was broken up. These engines were built to work between Portsmouth and London, and used to take the fast morning express.

The next engines to be built at Brighton were some large six-wheels coupled goods, numbered 224 to 227, *(Fig. 90)*. The cylinders were 17 inches in diameter with a stroke of 24 inches, and were placed at an inclination of 1 3/16 in 12, with their centres 2 feet 7 inches apart. The steam ports were 1⅜ inches wide by 14 inches long, and the exhaust ports 3 1/16 inches by 14 inches. The diameter of the wheels was 5 feet 1 inch, the leading and driving centres being 7 feet 8 inches apart, and the driving and trailing 8 feet 3 inches, giving a total wheel-base of 15 feet 11 inches. The leading overhang was 5 feet 3 inches and the trailing 3 feet 7 inches. The boiler was 4 feet 2 inches in diameter and 10 feet 6½ inches long, the centre line being 6 feet 7 inches above the rails. The steam dome was 2 feet 3 inches in diameter and 3 feet 1½ inches high. The boiler contained 154 tubes of 2 inch diameter and 10 feet 11½ inches long. The heating surface of the tubes was 870 sq. ft, and that of the firebox 106 sq. ft, giving a total of

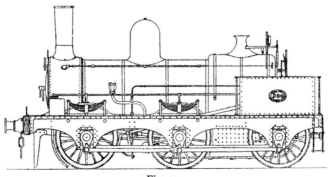

Fig. 90.

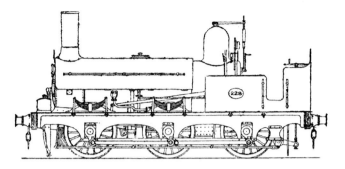

Fig. 91.

976 sq. ft. The length of the outside firebox was 5 feet 4 inches, the width 4 feet 1 inch, and the height inside 5 feet 6½ inches. These engines had double-beat regulators, one pump and one injector, and Allan's straight link motion. Their weight was 39 tons 6 cwt, of which the leading wheels carried 13 tons 2 cwt, the drivers 14 tons 2 cwt, and the trailers 12 tons 2 cwt. Nos 224 and 225 were built in June, 1866, and Nos 226 and 227 the next month, afterwards renumbered 390 to 393. No. 393 was scrapped in April, 1895, but Nos 390 to 392 were again re-numbered 465 to 467, and were scrapped in October, 1895. No. 466 finally was used for pumping purposes at Three Bridges. In March, 1898, it was renumbered 514, and again in May, 1899, to No. 614, and was broken up in July, 1901.

In August, 1866, two six-wheels coupled saddle tank engines were built at Brighton, numbered 228 and 229, *(Fig. 91)*. The cylinders, which were 17 inches in diameter by 24 inch stroke, were placed at an angle of 7⅝ inches in 7 feet, the centres being 2 feet 7 inches apart. The steam ports were 1⅜ inches by 14 inches, and the exhaust ports 3 1/16 inches by 14 inches, the diameter of the blast pipe orifice being 5 inches. The diameter of the wheels was 4 feet 9 inches, whilst the wheelbase was 14 feet 4 inches, equally divided. The boiler was 4 feet 1 inch in diameter and 10 feet long, the centre line being 6 feet 3 inches above the rail level; it contained 180 tubes 2 inches in diameter and 10 feet 4¾ inches long. The heating surface of the tubes was 900 sq. ft, and that of the firebox 80 sq. ft. The length of the outside firebox was 4 feet 3 inches and the breadth 4 feet 1 inch, the height being 4 feet 3 inches. The steam dome was over the firebox, and was 2 feet in diameter by 2 feet 10½ inches high; it contained a double-beat regulator. There was a man-hole under the tank, on the middle ring of the boiler. The capacity of the saddle tank, which was quite flat on the top, was 900 gallons. The weight on the driving wheels was 15 tons. The boiler was fed by one pump and one injector. Stephenson link motion was provided, having the weigh-bar shaft above. The brake was applied to the wheels by a screw and wheel, placed horizontally. These engines were re-numbered 351 and 353 in March, 1879, and in December, 1882, were sold to the Alexandra Dock Co., of Newport, Mon.

In October, 1866, a four-wheels coupled tank engine, No. 230, *(Fig. 92)*, was built at Brighton for working the Midhurst branch. The cylinders were 16 inches in diameter and 20 inch stroke, the centres being 2 feet 5 inches apart, and the inclination 5⅜ inches in 4 feet. The steam ports were 1⅜ inches by 15 inches, and the exhaust 3 inches by 15 inches. The travel of the valves, which were worked by the Stephenson link motion, was 4½ inches, the lead 3/16 inch, and the lap ⅞ inch. The leading and driving wheels were 5 feet in diameter and the trailing 3 feet 9 inches. The wheelbase was 14 feet 4 inches, the leading and driving centres being 6 feet 3 inches apart, and the driving and trailing 8 feet 1 inch; the leading overhang was 5 feet 8½ inches and the trailing 5 feet 2 inches. The external diameter of the boiler was 3 feet 11 inches, and its length 9 feet 9 inches; it contained 162 tubes 2 inches in diameter and 10 feet 2½ inches long, its centre line was 6 feet 6 inches above the rails. The heating surface of the tubes was 840 sq. ft, and of the firebox 90 sq. ft. The steam dome was 2 feet in diameter and 3 feet 4 inches high, and contained a double-beat regulator. The length of the outside firebox was 4 feet 8 inches, the breadth 4 feet 1 inch, and the length 5 feet 3¾ inches. The boiler was fed by one pump and one injector. The capacity of the side tanks was 580 gallons, and of the tank under the bunker 530 gallons. The bunker carried 1¼ tons of coke. The weight of this engine was 35 tons 9 cwt, of which 11 tons 18 cwt were on the leaders, 12 tons 3 cwt on the drivers, and 11 tons 8 cwt on the trailers. This engine was never re-numbered, and in June, 1881, it was scrapped.

For working on the South London line, Craven designed and constructed at Brighton in October, 1866, a four-wheels coupled bogie tank engine, very similar to the No. 235 class on the SER, built about the same time for the Charing Cross and Cannon Street service, *(Fig. 93)*. Numbered 231, it had cylinders 16-inches in diameter with a stroke of 20 inches, the centres being 2 feet 5 inches apart. The steam ports were 1⅜ inches wide by 14 inches long, and the exhaust ports 3 inches by 14 inches. The leading and driving wheels were 5 feet in diameter and the bogie wheels 3 feet 5 inches. The leading overhang

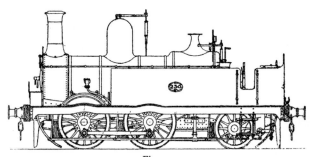

Fig. 92.

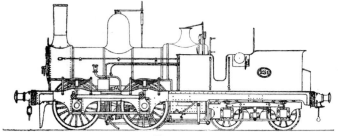

Fig. 93.

was 4 feet 10 inches, the coupled wheel-base was 7 feet 3 inches, from the driving centre to the centre of the bogie was 11 feet 9 inches, whilst the centres of the bogie wheels were 6 feet apart, making a total wheel-base of 22 feet. From the bogie centre to the front of the trailing buffer plank was 3 feet 4 inches. The boiler was of the usual design, being 3 feet 11 inches in diameter outside the largest ring and 9 feet 9 inches long, the height of the centre line above the rail level being 6 feet 5½ inches. The steam dome was 3 feet 2 inches high and 2 feet in diameter. The length of the firebox outside was 4 feet 8 inches, the breadth 4 feet, and the height inside 5 feet 10 inches. The weight on the coupled wheels was 26 tons, 12 tons being on the leading and 14 tons on the driving wheels. In May, 1884, this locomotive was re-numbered 466, and in June, 1886, was broken up. It had the usual Stroudley alterations, and was one of the first to be fitted with the Westinghouse brake. Latterly it worked on the Eastbourne branch.

It was followed by four single express locomotives, built for the Portsmouth and London service, numbered 232–235; *(Fig. 94)*. Nos 232 and 233 were constructed in October, 1866, and Nos 234 and 235 in November of that year. The cylinders, placed horizontally with the centres 2 feet 5 inches apart, were 16 inches in diameter with a piston stroke of 20 inches. The steam ports were 1⅛ inches by 14 inches, and the exhaust ports 3 inches by 14 inches. The wheelbase was 14 feet 6 inches, equally divided, the leading overhang being 4 feet 9 inches and the trailing 3 feet 1 inch. The carrying wheels were 4 feet in diameter and the drivers 6 feet. The boiler was 3 feet 11 inches in diameter, its length being 9 feet 9 inches; the height of the centre line above the rail level was 6 feet 5⅝ inches. The dome was 1 foot 11 inches in diameter and 3 feet 2 inches high. Length of the outside firebox was 4 feet 8 inches, the breadth 4 feet, and the height inside 5 feet 3¾ inches. The valves were actuated by the Stephenson link motion, having the weigh-bar shaft below. The weight was thus distributed: 10 tons 9 cwt on the leading wheels, 12 tons 13 cwt on the drivers, and 7 tons 14 cwt on the trailers, making 30 tons 16 cwt. No. 232 was in June, 1884, re-numbered 485, and in October of the same year was broken up. No. 233 was named *Horsham* by Stroudley, and in August, 1882, re-numbered 487; in February, 1883, it was sold to the West Lancashire Railway Company, by whom it was numbered 6. No. 234 was in April, 1881, re-numbered 474, and scrapped in October, 1884. No. 235 was re-numbered 475, and named by Stroudley *Dorking*; it was sold with No. 233 to

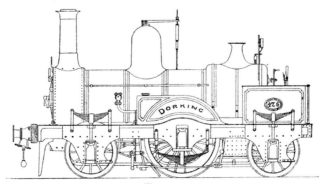

Fig. 94.

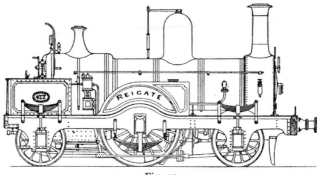

Fig. 95.

the West Lancashire Company, by whom it was numbered 5.

These were followed by six express locomotives constructed by Nasmyth, Wilson & Co., of Bridgewater Foundry, Patricroft, near Manchester, the maker's Nos being 114–119. They are illustrated by *Fig. 95*, and were numbered 236–241. Nos 236 and 237 came in April, 1867, Nos 238 and 239 in May, and Nos 240 and 241 in June of the same year. Nos 236, 238 and 240 had 16 inch cylinders, but those of the other thee were 16½ inches in diameter, the stroke of the pistons being 22 inches in each case. The diameter of the blast pipe orifice was 4½ inches. The wheelbase was 16 feet 2 inches, equally divided. The leading wheels were 4 feet 3 inches in diameter, the driving 6 feet 6 inches, and the trailing 4 feet. The diameter of the boiler was 3 feet 11¼ inches and the length 11 feet. It contained 153 tubes 2 inches in diameter and 11 feet 5 inches long. The length of the inside firebox was 4 feet 6⅞ inches, the breadth 3 feet 3⅛ inches, and the height 5 feet 5½ inches. The heating surface of the firebox was 124 sq. ft, and of the tubes 960 sq. ft, giving a total of 1,084 sq. ft. The firegrate area was 15.2 sq. ft and the working pressure 140 lbs per sq. in. The average weight of these engines was about 32 tons 14 cwt. Stroudley named them *Arundel, Reigate, Shoreham, Polegate, St Leonards* and *Eastbourne* respectively, and in September, 1881, they were re-numbered 476 to 481. No. 477, *Reigate*, was the first of these singles to be broken up, this being in June, 1888. It was followe din December, 1889, by No. 476, *Arundel*, and No. 481, *Eastbourne*. Nos 478, 479 and 480 all went together in June, 1894.

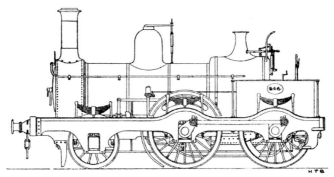

Fig. 96.

The next class of engine was ordered from Dübs & Co., of Glasgow, and consisted of six four-wheels coupled express locomotives, the maker's Nos being 131–136, and the railway company's 242–247; *(Fig. 96)*. No. 242 commenced work in March, 1867, and the remainder in April. Owing to a dispute between the company and the builders, the former would not received them at Lillie Bridge sidings, West Brompton, when brought there by the LNWR, the result being that they remained shunted in the sidings for twelve months. At the end of that time they were minus many brass fittings, etc., and having been exposed to the weather and the atmosphere of London without being cleaned, the state they had got into can be imagined. The cylinders, placed horizontally with centres 2 feet 5 inches apart, were 16 inches in diameter with a stroke of 20 inches. The steam ports were 1¼ by 14 inches and the exhaust ports 3¼ inches by 14 inches. From leading to driving and from driving to trailing centres were each 7 feet 6 inches, making a total wheelbase of 15 feet, the overhang at the leading end being 5 feet 3 inches and at the trailing end 4 feet. The leading wheels were 4 feet in diameter and the driving and trailing 6 feet. The diameter of the boiler was 4 feet, the length 10 feet 6 inches, and the height of the centre line above the rail level 6 feet 5 inches; it contained 156 2 inch tubes. The steam dome was 2 feet 1 inch in diameter and 2 feet 9½ inches high. The length of the outside firebox was 4 feet 10 inches, the breadth 4 feet, and the height inside 5 feet 3 inches. These engines weighed 33 tons 3 cwt, having 10 tons 17 cwt on the leading, 12 tons 19 cwt on the driving, and 9 tons 7 cwt on the trailing wheels. In September, 1881, they were re-numbered 457–462. The boiler centre of 457 was raised about 3 inches higher by Stroudley, who intended to rebuild these engines, but afterwards considered too costly the project was abandoned. No. 457 was scrapped in June, 1893; No. 458 in December, 1891; No. 459 in October, 1895; No. 460 in June, 1894; No. 461 in April, 1895, and No. 462 in November, 1894.

In November, 1867, Craven constructed at the Brighton two large goods engines with Cudworth's fireboxes. They were both stationed at New Cross, and one of them regularly worked the fast 5.05 p.m. goods train from Willow Walk to Newhaven. They were numbered 190 and 191, *(Fig. 97)*. Cylinders were 17 inches in diameter, and the stroke of the pistons was 24 inches, the centres being placed 2 feet 7 inches apart. The steam ports were 1½ inches wide and 16 inches long, and the exhaust ports 4 inches

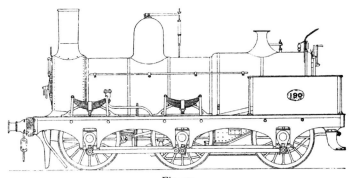

Fig. 97.

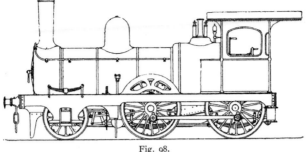

Fig. 98.

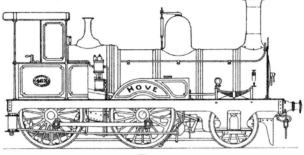

Fig. 99.

by 16 inches. The wheel-base was 15 feet 11 inches, being 7 feet 8 inches between the leading and driving centres, and 8 feet 3 inches between the driving and trailing; the overhang at the leading end was 5 feet 2 inches, and at the trailing 5 feet. The diameter of the wheels on the tread was 5 feet 2 inches. The boiler was made in three rings, the longitudinal and circumferential seams being butt-jointed, and was 4 feet 3 inches in external diameter, and 10 feet 6 inches long. The front tube plate was connected to the back plate of the firebox casing by six longitudinal stays. The steam dome was 2 feet 1 inch in diameter outside, and 2 feet 10 inches in height, steam being admitted to the cylinders by a double-beat regulator. The length of the firebox casing was 8 feet, the breadth 4 feet 1 inch, and the height inside was 5 feet 3½ inches at the front end and 3 feet at the back. The inside firebox was secured to the firebox casing by 16 girder stays placed athwart the box. The boiler had 192 brass tubes 2 inches in diameter and 10 feet 11 inches in length, and its centre line was 6 feet 8 inches above the rail level. Slide valves were worked by the Stephenson type of link motion, the weigh-bar shaft being underneath. These engines had both inside and outside frames, both of which were of plate iron, the driving wheels having inside and also outside bearings. Originally one spring balance safety valve was on the dome, and two over the firebox casing; but Stroudley, as he did with the majority of the Craven engines, took the one off the dome, and at a later period removed those from the firebox and put two on the dome. The average weight of these locos was 38 tons 4 cwt, distributed as: On the leading wheels 12 tons 8 cwt, on the driving 13 tons 6 cwt, and on the trailing 12 tons 10 cwt. Never re-numbered, No. 190 was scrapped in January, 1885, No. 191 in August, 1884.

In February, 1868, a four-wheels coupled tender engine was put to work on the main line. It was numbered, 248, *(Figs 98 and 99)*. It was built by Kitson & Co., of the Airedale Foundry, Leeds, and exhibited by that firm at the Paris Exhibition of 1867, where it was awarded a gold medal. It was numbered 1423 in the books of the makers, and when purchased was painted a light green lined with red, the framing being chocolate. From *Fig. 98* it will be seen that this engine had inside bearings to the coupled wheels, and outside to the leading. The unusually large cab, which had a decidedly foreign appearance, is very noticeable, and also the very low centre of the boiler. The cylinders, which were horizontal, were 16 inches in diameter, with a stroke of 22 inches, the centres being placed 2 feet 5 inches apart. The steam ports were 1¼ inches wide by 13 inches long, the exhaust ports being 3 inches by 13 inches. The leading wheels were 4 feet in diameter, and the driving and trailing 5 feet 6 inches. The wheelbase was 15 feet 6 inches, being 7 feet 9 inches between both the leading and driving and the coupled wheel centres. The overhang at the leading end was 4 feet 2⅝ inches, the leading buffer beam being of 1 inch plate iron, and the trailing overhang was 3 feet 8¾ inches. A curious point about this engine was that the throw of the outside coupling rods was 10¼ inches. The barrel of the boiler was made in three rings, being butt-jointed both in the circumferential and longitudinal seams. The diameter of the barrel outside was 3 feet 10 inches, and the length 10 feet 6 inches, the centre line being only 6 feet 1¼ inches above the rail level. The external diameter of the dome was 1 foot 10 inches, and the height 2 feet 9 inches. The length of the outside firebox was 4 feet 9 inches; the breadth 4 feet 1¾ inches, and the height inside 4 feet 10¾ inches. The boiler contained 140 tubes of 2 inch diameter, and about 10 feet 11 inches long. The heating surface of the tubes was 790 sq. ft, and of the firebox 82 sq. ft, giving a total of 872 sq. ft. The firegrate area was 14.5 sq. ft, and the diameter of the blast pipe orifice 4½ inches. Slide valves were actuated by the Stephenson link motion, having the weigh-bar shaft underneath, and reversed by means of a wheel and screw placed on the right-hand side of the engine, in which position it remained until the engine was broken up, this being the only instance in which Stroudley did not remove it to the left-hand side. The steam dome contained a vertical slide valve regulator. The dome casing was of polished brass, as were also the safety valve columns and the hand rails. The boiler was fed by two injectors, and the safety valves, which were Naylor's patent, were mounted over the firebox. The face of the driving side rod splasher was of brass, on which was engraved the maker's name and the date. The weight of this locomotive was 31 tons 6 cwt, the leading wheels carrying 9 tons 8 cwt, the driving 11 tons 16 cwt, and the trailing 10 tons 2 cwt. The sandbox was originally below the running-plate, but as seen from *Fig. 99* Stroudley put on the usual combined sandbox and splasher, and named the engine *Hove*. The upper portion of the cab side and roof was also new; and, further, Stroudley fitted the chimney with a copper cap, and took Naylor's valves off the firebox, two spring balance safety valves being substituted. These last were later put over the dome. The position of the clack boxes was altered, the brass safety valve funnel which was placed over the firebox was originally on the 'Jenny Lind' engine *Lewes (Fig. 14)*. In later years this engine was fitted with the Westinghouse automatic brake; in October, 1881, it was re-numbered 463, and in November, 1893, scrapped.

In March, 1868, a six-wheels coupled wing-tank engine for shunting at Willow Walk goods yard was constructed at Brighton from the designs of Craven. It was numbered 52, *(Fig. 100)*. The cylinders were 16 inches in diameter with a stroke of 24 inches, and were placed 2 feet 7 inches between centres. The steam ports were 1½ inches wide by 16 inches deep, the exhaust ports being 4 inches by 16 inches. The diameter of the wheels was 4 feet 9 inches. The overhang at the leading end was 5 feet 3 inches, and at the trailing end 5 feet 6½ inches. The wheel-base was 15 feet, being 7 feet 8 inches between the leading and driving centres, and 7 feet 4 inches between the driving and trailing. The boiler, which was of the usual design, was 3 feet 11 inches in diameter outside the largest ring, the length of the barrel being 10 feet 6 inches It contained 155 brass tubes 2 inches in diameter and 10 feet 11 inches long, and its centre line was 6 feet 4 inches above the rail level. The diameter of the steam dome, outside, was 2 feet 2 inches, steam being admitted to the cylinders by a double-beat regulator, which it retained up to the last; this is a remarkable fact, as all Craven's engines with this exception were fitted with the Stroudley regulator when they were brought into the shop. The length of the outside firebox was 4 feet 1 inch, the breadth 4 feet 1 inch, and the height inside 5 feet 1 inch. Two spring balance safety valves were placed over the firebox casing, these also remaining up to the finish, instead of being removed to the dome as was usually done, and the old Craven chimney was also retained. The boiler was fed by one pump and one injector, the injector clack box being on the firebox front by the gauge columns. As will be seen from the illustration, this engine had both outside and inside frames; these were of plate iron running the whole length of the engine, the driving wheels having four bearings. The wing tanks were 13 feet 3 inches long, and were supported on vertical iron plates. A tank was also under the footplate, behind the trailing axle, and would carry 183 gallons. The slide valves were actuated by the Stephenson link motion, having the weigh-bar shaft below. In September, 1875, it was re-numbered 269, and in April, 1880, it was again altered to 395, the number being painted on the side sheets. In later years this engine was used as pilot in the loco yard at Brighton, previous to which it was pilot at Newhaven goods wharf for a number of years. In October, 1893, it was sent up to Horley to be broken up, and, excepting little No. 27, which will be described next, was the last of Craven's tank engines. The weight of this engine was 37 tons 1 cwt, distributed as follows: 12 tons 3 cwt on the leading wheels, 13 tons 3 cwt on the drivers, and 11 tons 15 cwt on the trailers.

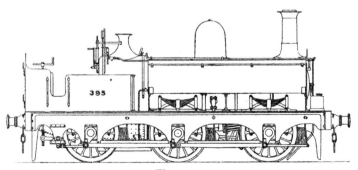

Fig. 100.

For working the branch line from Polegate to Hailsham, a distance of 2 miles 76 chains, with a train made up of three coaches, Craven constructed at Brighton in July, 1868, a neat little saddle tank locomotive, having only four wheels and outside cylinders. Numbered 27, it is illustrated by *Figs 101, 102 and 103*, showing it in various styles after its several alterations, its original condition being as in *Fig. 101*. The cylinders were 9 inches in diameter and the stroke of the pistons 14 inches, the centres being placed 6 feet 4¼ inches apart. The steam ports were 1½ inches wide and 7 inches deep, the exhaust ports being 2½ inches by 7 inches. The diameter of the wheels was 3 feet 2 inches and the wheel-base 6 feet. The overhang at the leading end was 4 feet 10½ inches, and 6 feet 10½ inches at the trailing end. The boiler barrel was made in two rings owing to its short length, and was butt-jointed both longitudinally and horizontally, the diameter outside being 2 feet 11 inches and the length 8 feet. It contained 63 tubes, 2 inches in diameter and 8 feet 5 inches in length, the centre line of the barrel being only 4 feet 9 inches above the rails, whilst the height of the chimney

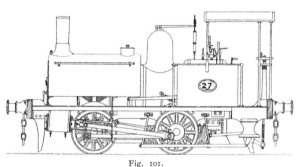

Fig. 101.

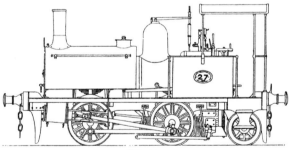

Fig. 102.

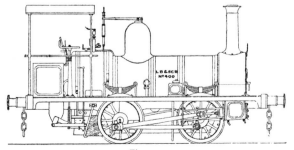

Fig. 103.

top was 9 feet 9 inches. The length of the outside firebox was 2 feet 10⅞ inches, the height inside the copper box 3 feet 9 inches and the width 3 feet 7⅞ inches. The steam dome was 1 foot 9 inches in diameter and 2 feet 3 inches in height; two spring balance safety valves being fixed upon it. The capacity of the saddle tank was 320 gallons. The coke boxes carried 15 cwt and were at the sides of the footplate. The heating surface of the tubes was 274 sq. ft and that of the firebox 44 sq. ft, giving a total of 318 sq. ft. The wheels were of chilled cast iron, the tyres and centres being cast in one piece, with spokes of + section. The slide valves were worked by the Stephenson link motion, having the weigh-bar shaft above. It will be seen from the illustration that this engine had a cab, but there was no weather board. After it had been working on the branch a short time it was brought back to Brighton to have a pair of trailing wheels added, owing to its pitching badly when running, especially bunker first. The diameter of the trailing wheels was 2 feet 4 inches, the brake column being then shifted from the back to the side of the engine and a weather-board fitted up. After alteration this engine weighed 19 tons 10 cwt, the leading wheels carrying 7 tons 8 cwt, the driving wheels 8 tons 14 cwt and the trailing 3 tons 8 cwt. The trailing overhang was then 1 foot 7 inches, the distance between the driving and trailing centres being 5 feet 3½ inches. This engine was then sent down to work the Littlehampton and Arundel branch, which it continued to do satisfactorily for some time. When, however, Stroudley required a small engine for shunting in the works at Brighton, this engine was selected as being most suitable for the purpose. To enable the engine to take the sharp curves in the works with greater ease the trailing wheels were removed, and it was once more a four-wheeled engine. The saddle tank was removed and a well tank substituted, which was placed under the boiler barrel between the leading and driving wheels. The height of the chimney was increased 3 inches, the coal bunker was on the right-hand side only, and only one brake block was provided to each driving wheel instead of two as formerly. The weather-board was removed, the cab roof altered and a new angle iron fitted outside under the footplate. In March, 1879, this engine was sent to Earlswood for shunting at Redhill goods yard, where it worked for about sixteen years, when it was sent to Tunbridge Wells for pumping purposes. In April, 1896, it was sent to Horley to be scrapped. When the present No. 27 was built this engine was re-numbered 400. When stationed at Earlswood it used to go to Three Bridges shed every week end to have the boiler washed out. It was formerly painted the standard colour for passenger engines, with the lettering LBSCR in gold and No. 400 underneath; but latterly it was painted the goods engine green with the writing in yellow. When this locomotive was altered by Stroudley larger wheels were put in, their diameter being 3 feet 5½ inches on tread, with tyres 2½ inches thick. The weight on the leading wheels was then 9 tons 10 cwt and on the driving 5 tons 12 cwt, giving a total of 15 tons 2 cwt. The boiler was always fed by one Giffard injector.

In October, 1868, six goods engines of the ordinary six-wheels coupled type were put to work on the main line. These were built by Slaughter & Co., of the Avonside Engine Works, Bristol, the builders' numbers being 744–749, and those of the railway company 249–254. *Fig. 104*. The cylinders were 17 inches in diameter, with a stroke of 24 inches, the centres being 2 feet 7 inches apart. The steam ports were 1½ inches wide

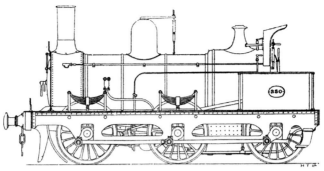

Fig. 104.

and 1 foot 4 inches long, the exhaust ports being 3½ inches by 1 foot 4 inches. Diameter of the wheels on tread was 5 feet, and the wheelbase was 15 feet 11 inches, being 7 feet 8 inches between the leading and driving centres and 8 feet 3 inches between the driving and trailing. From the front buffer beam to the leading centre was 5 feet 2½ inches and the trailing overhang was 3 feet 7 inches. The boiler was butt jointed both in the circumferential and longitudinal seams, and double rivetted, the external diameter being 4 feet 3 inches, the length 10 feet 6 inches and the height of the centre line above the rail level 6 feet 8 inches. The steam dome was 2 feet 1 inch in diameter and 2 feet 10 inches in height; it contained a vertical slide valve regulator. The length of the firebox casing was 5 feet 4 inches, the breadth 4 feet 1 inch and the height of the inside firebox 5 feet 7½ inches. The front tube plate was connected to the back plate of the firebox casing by nine longitudinal stays and the inside box to the outer shell by eight girder stays. The heating surface of the tubes was 990 sq. ft, and that of the firebox 100 sq. ft, giving a total of 1,090 sq. ft. The boiler was fed by one injector and two pumps. The valves were worked by the Stephenson link motion, having the weigh-bar shaft below. As will be seen from the illustration, these engines had outside frames, very solid and deep, and having large oval openings between the coupled wheels. The inside frames ran the whole length of the engine. When new there was a strip of brass running round the face of the leading and driving splashers, but these were painted over by Stroudley, although within recent years one or two of them had the brass polished by some enterprising firemen. These engines originally had one spring balance over the dome and two over the firebox, but they underwent the usual changes. In October, 1881, they were re-numbered 468–473. Just previously to Stroudley's death, in 1889, No. 468 was fitted with injectors, and at a later date No. 471 was also. For a number of years No. 468 had a copper top to its chimney, which it retained to the last. The weight of these engines was about 35 tons 17 cwt each, distributed as follows: 12 tons 12 cwt on the leading wheels, 13 tons 7 cwt on the driving and 9 tons 18 cwt on the trailing. The sandboxes were originally between the frames in front of the leading wheels, but were afterwards put above the framing, some in front of the smokebox and others between the smokebox front plate and the leading splasher. When new the dome casings and handrails were of polished brass; on one or two the latter were replaced by iron handrails. These goods engines were the last of Craven's design, and were the finest. The capacity of the tender tanks was 2,000 gallons. In November, 1894, Nos 469 and 473 were sent to Horley to be scrapped, No. 472 went in April, 1896, and Nos 470 and 471 in September, 1896. No. 468 was again renumbered in March, 1898, 515, and was scrapped in January, 1899.

The last two single express engines built by Craven were Nos 255 and 256. Both engines were put to work in October, 1868, *(Figs 105 and 106)*. The former shows them in their original condition, the latter as altered by Stroudley. The cylinders were 17 inches in diameter and the stroke of the pistons 22 inches; they were placed horizontally, with the centres 2 feet 6 inches apart. From the centre of the cylinders to the centre of the driving axle was 9 feet 10⅞ inches. The steam ports were 1½ inches wide and 16¾ inches long, the exhaust ports being 4 inches by 16¾ inches. The leading wheels were 4 feet 3 inches in diameter, the driving wheels 6 feet 6 inches and the trailers 4 feet. Total wheelbase was 16 feet 2 inches, equally divided between each pair of centres. The boiler was 3 feet 11 inches in diameter by 11 feet long, the centre line being 6 feet 8 inches from rails; the outside firebox was 5 feet 3 inches long, the breadth being 4 feet and height inside 5 feet 5½ inches. One spring balance safety valve was on the dome and two over the firebox. The link motion was of the ordinary Stephenson type, having the weigh-bar shaft below. The boiler was fed by one pump and one injector. In their original state it will be seen that the sandboxes were connected to the driving splashers and the cabs had no sides. The splashers had double brass bands round the circumference. Under Stroudley's *régime* these engines underwent the usual alterations. They were both fitted with the Westinghouse brake and named *Hastings* and *Victoria* respectively. The former was stationed at New Cross and the latter at Battersea. In October, 1881, they were re-numbered 482 and 483, No. 482 having the number painted on in gold while No. 483 had a standard number plate. In June, 1888, No. 482 was sent to Horley to be scrapped, No. 483 followed in 1891.

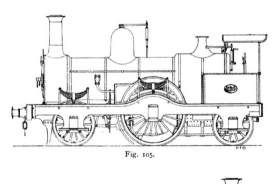

Fig. 105.

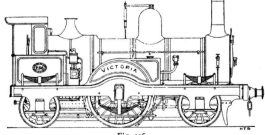

Fig. 106.

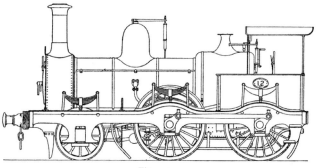

Fig. 107.

In November, 1868, Craven built a four-wheels coupled express engine, *(Fig. 107),* numbered 12. When new this engine was stationed at New Cross. The cylinders were 16 inches in diameter, with a piston stroke of 20 inches. The diameter of the leading wheels was 4 feet and of the coupled wheels 6 feet. Wheelbase was 15 feet 4 inches equally divided. The boiler was of the usual design, 3 feet 11 inches in diameter outside and 10 feet 6 inches long, the centre line from rail level being 6 feet 5 inches. The boiler contained 156 2 inch tubes. One spring balance was on the dome and two on the firebox. The firebox was 4 feet 10 inches long and in breadth 4 feet 0½ inches, the height inside being 5 feet 2½ inches. The boiler was fed by one injector and one pump. The valve gear was the ordinary link motion. The frame plates, both inside and outside, were of iron, and extended the whole length of the engine, the driving wheels having both inside and outside bearings. This engine was one of the first to have steel tyres. A spark arrester was placed inside the smokebox, fixed upon the top of the blast pipe and tapering outwards to the base of the chimney. The weight of this engine was 32 tons 10 cwt, distributed as follows: 10 tons 9 cwt on the leaders, 12 tons 13 cwt on the drivers and 9 tons 8 cwt on the trailers. This was the first coupled express engine to have a cab. Sides were afterwards added to it by Stroudley, and afforded better protection from the weather. It was first re-numbered 124, but in July, 1879, was again re-numbered 371, and in April, 1886, it was sent up to Horley to be scrapped.

Next came a small four-coupled passenger engine, also built in November, 1868.

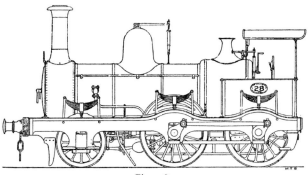

Fig. 108.

It was numbered 28, *(Fig. 108)*. The cylinders, horizontal, were 16 inch diameter by 20 inch stroke. The leading wheels were 4 foot diameter and the coupled wheels 5 feet 6 inches. The wheelbase was 14 feet 8 inches, being 7 feet 8 inches between the leading and driving wheels and 7 feet between the coupled wheels. The boiler was 3 feet 11 inches in outside diameter and 10 feet 6 inches long, containing 162 tubes 2 inches in diameter; its centre line was 6 feet 2 inches from the rail level. The outside firebox was 4 feet 1⅝ inches long and 4 feet 2½ inches wide. This was an unusually wide box and its height inside was 4 feet 10½ inches. The boiler was fed by one pump and one injector. One safety valve was on the dome and two over the firebox. Like No. 12 and others, it had the safety valves altered when it underwent the usual Stroudley modifications. The motion and other details were similar to those of No. 12, including the conical spark arrester in the smokebox. From the illustration it will be seen that it had outside frames, but it also had inside ones, with bearings inside and out for the driving wheels. As in Craven's earlier engines, the inside frames stopped at the front of the firebox casing, to which they were attached by a cast-iron bracket, thus enabling a wide firebox to be got in. The frames commenced again at the back plate and continued up to the trailing buffer beam. The weight of this engine was 31 tons, of which 10 tons 11 cwt was on the leading wheels, 12 tons 9 cwt on the driving and 8 tons on the trailing. It was first re-numbered 283, but in September, 1879, was again re-numbered 377, and in August, 1884, it was sent up to Horley to be scrapped. When No. 28 came out new it was sent to Dorking, but after a short time was transferred to New Cross and put to work on the main line. Being found unsuitable for this work, it was put on to local service and ran between London Bridge, Victoria, and Croydon.

In December, 1868, Craven built at Brighton a heavy six-wheels coupled wing-tank engine for shunting at the Willow Walk goods yard and piloting heavy trains up the New Cross bank, which has a gradient of 1 in 100. It was numbered 58, *(Fig. 109)*. The cylinders were 16 inch diameter, and were inclined, the stroke of the pistons being 24 inches and the centres 2 feet 7 inches apart. The steam ports were 1⅜ inches in width and 1 foot 3¾ inches deep, the exhaust ports being 4 inches by 1 foot 3¾ inches. The diameter of the wheels on tread was 4 feet 9 inches. Wheelbase was 15 feet 4 inches equally divided, the leading overhang being 5 feet 3 inches and the trailing 5 feet 6½ inches. The boiler barrel was 4 feet 3 inches in diameter outside and 10 feet

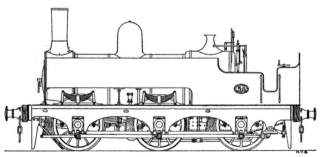

Fig. 109.

6 inches long. It was parallel and butt-jointed circumferentially and longitudinally, with its centre line 6 feet 6 inches above the rail level. The steam dome was 2 feet 1 inch in diameter. The length of the firebox casing was 4 feet 6 inches, the breadth 4 feet 1 inch and the height inside 5 feet 5 inches. The valve motion was of the Stephenson type, having the weigh-bar shaft below. This engine had inside and outside frames, the former running the whole length of the engine. A tank was also placed under the bunker behind the trailing axle. One pump and one injector supplied the boiler with water. There was no safety valve on the dome, but two over the firebox. The weight of the locomotive was 40 tons 17 cwt, and was distributed as follows: 12 tons on the leading axle, 16 tons on the driving and 12 tons 17 cwt on the trailing. A feature of this engine worthy of notice was that the wheels had only twelve spokes each. In October, 1875, it was re-numbered 273 and in April, 1880, again re-numbered 398. In September, 1885, it was sold to the Alexandra Dock Company, of Newport, Mon., by whom it was numbered 14.

In May, 1869, Kitson & Co., of Leeds, supplied the company with a small saddle tank engine having leading and driving wheels coupled; it was numbered 76, being No. 1553 in the books of the makers, *(Figs 110 and 111)*. The cylinders, which were inside, were 12 inches in diameter with a piston stroke of 18 inches, the centres being placed 2 feet 1 inch apart. The steam ports were 1 inch wide and 8 inches long, the exhaust ports being 2¼ inches by 8 inches. The diameter of the blast pipe orifice was 3 inches.

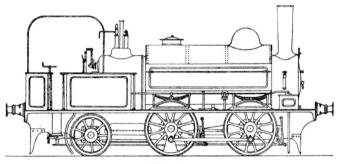

Fig. 110.

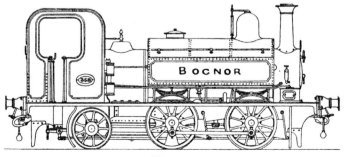

Fig. 111.

The diameter of the coupled wheels was 4 feet, and that of the trailing wheels 3 feet. The leading overhang was 5 feet 1 inch, from leading to driving centres was 6 feet, from driving to trailing 6 feet 2 inches, and the overhang at the trailing end was 5 feet 0½ inches. The barrel of the boiler was made in two rings, and was butt-jointed both longitudinally and circumferentially, its external diameter being 3 feet and its length 9 feet 7 inches, the height of the centre line above the rail level being only 5 feet 5½ inches. The firebox casing was raised and secured to the inside box by girder stays; its length was 3 feet 6 inches, and its width 3 feet 6 inches, the height of the inside box being 3 feet 11 inches. The steam dome, which was placed on the first ring of the barrel, was 1 foot 9 inches in diameter outside and 2 feet 6 inches high. The boiler contained 110 brass tubes 1¾ inches in diameter and 10 feet long, giving a heating surface of 452 sq. ft, which, with 52 sq. ft in the firebox, made a total heating surface of 504 sq. ft. The area of the firegrate was 9 sq. ft. The valve motion was of the Stephenson type, having the weigh-bar shaft above. The leading and driving springs were above the axleboxes, whilst the trailing wheels had only one spring, of the ordinary laminated type, but inverted and placed athwart the engine, taking its end bearings on the top of the axleboxes. The steam regulator was of the pull-out type, placed in the smokebox; steam was admitted through a bell-mouthed pipe in the dome. The safety valves were Naylor's patent, and the boiler was fed by one injector on the left-hand side and one pump on the right-hand side of the engine. The saddle tank had a half round top with flat sides, and carried 500 gallons of water. The weight of the engine was 17 tons 2 cwt, distributed as follows: 4 tons 15 cwt on the leading wheels, 5 tons 16 cwt on the driving, and 6 tons 11 cwt on the trailing. During Stroudley's *régime* sides were added to the cab, a copper-topped chimney fitted, and the safety valves replaced by the standard spring balances, besides minor alterations. It was also named *Bognor*, and in June, 1877, was re-numbered 358, but in August, 1886, was again altered to 496, and in April, 1895, it was scrapped. During its last few years this engine was running between Arundel and Littlehampton, being stationed at the latter town.

In October, 1869, the company purchased from Sharp, Stewart & Co., of Manchester, a small driving and trailing coupled side tank engine, *(Figs 112, 113 and 114)*. It was numbered 96 in the books of the railway company, and 1924 in those of the builders. The cylinders were inside, placed at an inclination of 1 in 9 with the centres 2 feet 3 inches apart, their diameter being 10 inches, and the stroke of the pistons 16 inches. The diameter of the blast pipe orifice was 3½ inches. The leading wheels were 2 feet 9 inches in diameter on tread, and the driving and trailing wheels 4 feet. The wheelbase was 11 feet, equally divided. The barrel of the boiler was in three rings, its external diameter being 2 feet 10 inches and its length 8 feet 6 inches; its centre line was very low, being only 5 feet 0½ inch above the rail level. The length of the firebox casing was only 2 feet 8 inches, the width 3 feet 7⅛ inches, and the height inside 3 feet 3 inches. The boiler contained 75 tubes of 1⅝ inch diameter. The heating surface of the tubes was 278.47 sq. ft, and of the firebox 33.75 sq. ft, giving a total of 312.22 sq. ft. The firebox casing was raised, and the crown of the inside box was carried by girder roof stays slung by links fixed in the manholes. From the illustration the engine appears to have a dome; this, however, is not one, but only a manhole, having a diameter of 1 foot

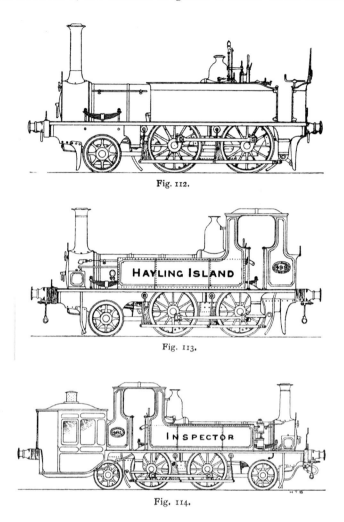

Fig. 112.

Fig. 113.

Fig. 114.

5¼ inches. The regulator was placed in the smokebox, and was of the pull-out type. In the smokebox there was a spark arrester tapering from the top of the blast pipe to the base of the chimney; the top of the latter was 11 feet from the rail level. The boiler was fed by one pump and one injector. The slide valves were worked by the Stephenson link motion, the weigh-bar shaft being above. The side tanks had a length of 9 feet 8 inches, and carried 300 gallons of water. When new this engine worked the Brighton and Kemp Town service, and was named *Kemp Town* by Stroudley; but when the new *Kemp Town*, No. 64, was built in June, 1874, No. 96 was taken off the branch and sent to Havant to work the Havant and South Hayling service; it was then renamed *Hayling Island*. From *Fig. 113* it will be seen that Stroudley put on a cab of his own design, and also fitted his standard chimney, added a dome, and increased the height of the casing. The engine was fitted with the Westinghouse brake, and various small alterations made, after which it weighed 19 tons 8 cwt, of which the leading wheels carried 6 tons, the driving 6 tons 8 cwt, and the trailing 7 tons. For the use of the engineer when inspecting permanent way, bridges, etc., Stroudley added a small saloon to the back of this engine, and an extra pair of trailing wheels; but owing to his sudden death in December, 1889, he did not see these alterations finished. The engine was then re-named *Inspector*; its number was first altered from 96 to 115; in June, 1877, it was again re-numbered 359, whilst in January, 1886, it was made 499; finally it was altered to 481 when it was named *Inspector*. When the cab was added the side tanks were renewed and made longer. A speaking tube communicated from the saloon to the driver, and it was also fitted with a speed indicator. When this engine ran it carried a white board with black horizontal stripes, similar to engines running over the new line from Croydon to Redhill. The locomotive was scrapped in January, 1899.

In December, 1869, Craven built a four-wheels coupled side tank engine for working the East London line. It was numbered 51, *(Fig. 115.* The cylinders, which were horizontal, were 15 inches in diameter, the stroke of the pistons being 20 inches, and the centres placed 2 feet 5 inches apart. The steam ports were 1⅛ inches wide and 1 foot 3 inches deep, the exhaust ports being 2⅛ inches wide by 1 foot 3 inches. The diameter of the leading wheels was 3 feet 6 inches and of the driving and trailing wheels 5 feet. The wheelbase was 15 feet, from leading to driving centres being 7 feet 10 inches, and from driving to trailing 7 feet 2 inches. The overhang at the leading end was 4 feet 11 inches, and at the trailing end 5 feet. The boiler was made in three rings, the external diameter being 3 feet 8 inches, and the length 10 feet 6 inches. The number of tubes was 158, their diameter being 1¾ inches. The length of the firebox casing was 4 feet 0⅛ inches, the width 4 feet 1¼ inches, and the height inside 4 feet 7½ inches. The steam dome was 1 foot 10 inches in diameter. The centre line of the boiler was 6 feet above the rail level. The heating surface of the firebox was 757 sq. ft, and that of the tubes 73 sq. ft, giving a total of 830 sq. ft. Length of the side tanks was 12 feet 4 inches, and the capacity 355 gallons of water. There was also a tank under the bunker, behind the trailing axle, which would carry 200 gallons, thus giving a total water capacity of 555 gallons. The driving and trailing springs were connected by a compensating beam. The valves were worked by the Stephenson link motion, having the weigh-bar shaft above. From the diagram it will be seen that the engine had outside bearings to the leading wheels, and inside to the driving and trailing. The inside frames ran the whole length of the engine, and the total length of the engine over the buffers was 32 feet

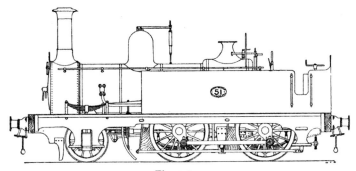

Fig. 115.

11 inches. One spring balance safety valve was on the dome, and two over the firebox casing. The weight of this engine was 34 tons 17 cwt, which was distributed as follows: On the leading wheels 9 tons 5 cwt, on the driving wheels 13 tons, and on the trailing wheels 12 tons 12 cwt. This locomotive finished work at Barnham Junction, to which place it went down from New Cross under its own steam to do duty as a pumping engine until it was scrapped in January, 1885. It was first re-numbered 132, but in September, 1878, it was again altered to 379.

This was followed in December, 1869, by another for the East London line of the same design but of slightly smaller dimensions, numbered 109. The cylinders and boiler were identical with those of No. 51, but the firebox was 4 inches shorter, the length outside being 3 feet 8⅛ inches. The wheels were the same also, but the wheelbase was reduced 4 inches, the centres being 7 feet 10 inches between leading and driving, and 6 feet 10 inches between driving and trailing. The side tank was also 4 inches shorter than in No. 51. The heating surface of the firebox was 68 sq. ft, and with the 757 sq. ft of the tubes, the total was 825 sq. ft. The boilers of both these locomotives were fed by one injector placed on the right-hand side of the engine, and one long-stroke pump on the left-hand side. In February, 1877, this engine was re-numbered 352, and in May, 1885, the boiler was sent to Deptford Wharf for supplying steam to the engines that worked the coal hoist, the framing, wheels, etc., being scrapped. These two locos were the last to be built at Brighton under the superintendence of John Chester Craven.

About the year 1866 the company let out a contract for six express locomotives of the single-wheel type, which was undertaken by Dodds & Son, of the Holmes Engine and Railway Works, Rotherham. This firm had also an order for some tank engines for a Spanish railway, but at that time the Carlist rising was in full force in Spain, the result being that the Spanish railway received their engines, but the locomotive builders did not get their dollars, and consequently went into liquidation, the works being closed never to be reopened as a locomotive building establishment. It thus happened that of the Brighton engines, out of the order for six, only two were built; numbered 127 and 128, *(Fig. 116)*. They were Nos 69 and 70 in the books of the makers. When No. 127 was finished it was sent to the Attercliffe running sheds (Midland Railway) near Sheffield, and remained there until it was brought to Brighton in 1871. No. 128 remained in the shops of Dodds unfinished for some time, the slide bars being

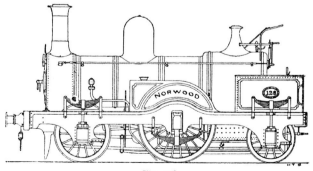

Fig. 116.

afterwards fitted up on arrival at Brighton, as well as the motion and connecting rods. No line of railway was connected to the works, these being situated on the banks of the River Don, and in order to remove the engines they had to be taken out with the help of jacks, drawn over some fields and roads, and finally placed in barges and taken up the river until they could be put on the railway. Unfortunately, few dimensions of these fine locomotives are to be obtained. The cylinders were 17 inches in diameter, and the stroke of the pistons 22 inches. The wheelbase was 16 feet, equally divided. The leading wheels were 4 feet 6 inches in diameter, the driving 6 feet 6 inches, and the trailing 4 feet. The weight on the leading wheels was 11 tons, on the driving 14 tons 16 cwt, and on the trailing 8 tons, making a total of 33 tons 16 cwt. When new they had one spring balance safety valve on the dome and two over the firebox, but these and the driving splashers and sand-boxes were altered by Stroudley like the majority of the others. In their original condition the sand-box was above the footplate, butting up to the splasher. They had the Stephenson link motion, the weigh-bar shaft being below. No. 127 was named *Norwood* by Stroudley, and No. 128 *Croydon*. In September, 1878, they were re-numbered 374 and 375 respectively. In December, 1892, No. 374 was sent to Horley to be scrapped, No. 375 having preceded it in June, 1888.

John Chester Craven

Having completed the history of the locomotives built under the superintendence of Craven, we may add a few notes on his life. John Chester Craven was the son of a millwright, and was born at Hunslet, near Leeds, on 11 September, 1813. At fourteen years of age he was apprenticed to the famous firm of Fenton, Murray & Jackson, of Leeds. In March, 1837, he left during the great strike and accompanied Telford, afterwards manager and partner in the firm of Carrett, Marshall & Co., to the Sun Foundry, Leeds. From there he went to Maudslay & Co., of Westminster, but after twelve months returned to Leeds to be foreman and afterwards manager for Todd, Kitson & Laird. He next joined Shepherd in the firm of Shepherd & Todd, of the Railway Foundry, Leeds (afterwards E. B. Wilson & Co.), where he was manager for three years, when he left to become locomotive foreman on the Manchester & Leeds Railway, now part of the Lancashire & Yorkshire. He left this company in May, 1845, and joined the locomotive department of the Eastern Counties Railway at Stratford; but in November, 1847, was appointed locomotive superintendent of the LBSCR, which post he held until his retirement at the close of the year 1869. Latterly his time was largely occupied with steamships, and in conjunction with others he started the Victoria Dock Engine Works. He was elected a member of the Institution of Civil Engineers on 4 December, 1866, and in 1871 he was made a town councillor in the Brighton Corporation, and in 1881 Alderman of Brighton. He died on 27 June, 1887.

Chapter 4

William Stroudley, 1879 – 1881

In March, 1871, the directors of the LBSCR appointed William Stroudley locomotive, carriage and wagon superintendent, that gentleman having occupied a similar position on the Highland Railway since January, 1866. The first two locomotives built by him were four-wheels coupled side tanks, numbered 18 and 21, and completed in December, 1871. The boilers of these engines were not new, being in fact taken out of the Stephenson engines Nos 198 and 204, which were converted into coupled engines and reboilered. The cylinders, which were inclined, were 16 inches in diameter, the stroke of the pistons being 20 inches and the centres 2 feet 5 inches apart. The diameter of the blast pipe orifice was 4 inches. Wheelbase was 14 feet, equally divided. The diameter of the leading and driving wheels was 5 feet and of the trailers 3 feet 8 inches. The overhang at the leading end was 4 feet 9½ inches and at the trailing end 6 feet 6 inches. By referring to *Fig. 117* the general outline of Stroudley's well-known standard tank engine will be noticed. The horn-blocks or axlebox guides of these locomotives were of rather peculiar design, having very large solid flanges prolonged on each side with the brackets for carrying the spring hangers cast solid with them. The centre line of the boiler was 6 feet 5¼ inches above the rail level. The height of the side tanks was 4 feet and the length 14 feet. Slide valves were worked by the Stephenson link motion, but the reversing gear was of rather unusual design. The reversing lever was pivoted on a centre carried by a bracket on the side of the tank, the upper end having a toothed sector into which the teeth of a pinion engaged; this latter was fast on the reversing wheel shaft. The wheel was of the marine steering type, having handles on the rim, and stood parallel to the centre line of the engine. The bottom end of the lever was connected in the usual manner to the reversing rod. No. 18 finished its career as pilot at Eastbourne, and No. 21 at Horsham. No. 18 was re-numbered by Stroudley 261, and in July, 1881, again altered to 373; it was scrapped in May, 1886. No. 21 was first re-numbered 265, and in October, 1881, was made 467. It was scrapped about 1888.

In December, 1871, two goods engines were turned out of the Brighton Works from the designs of Stroudley. From *Fig. 118* and the following particulars, it will be observed at a glance what powerful and up-to-date locomotives they are. For ordinary traffic they were at the time the largest goods engines in this country; they were numbered 84 and 85. The former commenced work on 3 December and the latter on 9 December, 1871. The cylinders, which were 17½ inches in diameter with a stroke of 26 inches, were both made in one casting, including the steam chest, steam and exhaust branches, and as the valve faces were placed underneath, the centres of the cylinders were capable of being brought very close together, namely, 2 feet 1 inch. They were placed at an angle of 1 in 11½, and had steam ports 1⅜ inches wide by 15 inches long, the exhaust ports being 2 inches by 15 inches. The blast pipe orifice was 4¾ inches in diameter and 13¼ inches above the centre of the boiler. The inclination of the slide valves was 1 in 16½. The engine frames, which are inside, are very deep throughout, being made of Monkbridge iron 1⅛ inches in thickness. The diameter of the wheels on tread is 5 feet. From the front of the leading buffer beam to the leading wheel centre is 5 feet 1¼ inches, the wheel-base is 15 feet 3 inches, being 7 feet 9 inches between the leading and driving centres, and 7 feet 6 inches between the driving and trailing centres, and the overhang at the trailing end is 4 feet 7¼ inches. The boiler is flush topped and of the telescopic type, single-rivetted at the circumferential joints, and double-rivetted with inside and outside butt strips at the longitudinal joints. The thickness of the plates is 9/16 inch, the external diameter of the boiler being 4 feet 3¾ inches at the front ring, 4 feet 4⅞ inches at the middle ring, and 4 feet 6 inches at the third ring. These engines do not appear to have such a large boiler owing to the centre line being rather low, namely, 6 feet 10½ inches. The barrel contained originally 311 brass tubes 1½ inches

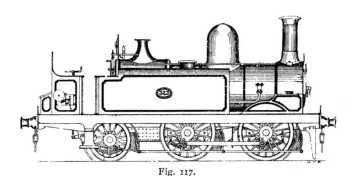

Fig. 117.

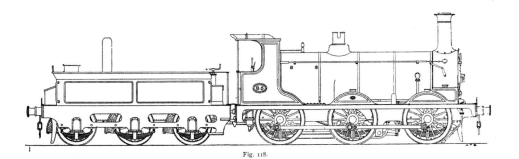

Fig. 118.

in diameter outside and 10 feet 11 inches in length, but there are now only 248 tubes of 1¾ inch diameter. The length of the firebox casing is 6 feet 2 3/16 inches, the width 4 feet 1 inch, and the depth below the centre line of the boiler at the front end 4 feet 11¾ inches; but at the back it is only 3 feet 1½ inches, so as to allow the trailing axle to come underneath, thereby keeping the wheel-base as short as possible, which was a leading characteristic of Stroudley's engines. The heating surface both of the 1½ inch and 1¾ inch tubes is 1,312 sq. ft, and that of the firebox 102, giving a total of 1,414 sq. ft. The area of the firegrate is 19.5 sq. ft, and the working pressure of the boiler 140 lbs per sq. in. When these locomotives were originally built they had no steam domes, and over the manhole was fixed Adams' patent safety valves. These latter were removed after a year or so, not having proved very successful; they would often seize, and then when the pressure accumulated very high would 'pop' off with a loud report. The tenders of these engines, which are on six wheels of 4 foot diameter, have outside frames and are very solidly built. The wheel-base is 12 feet, equally divided. The centres of the axlebox journals are 6 feet 3 inches apart, the tank being 16 feet long and 3 feet 4¼ inches to the top of the coping from the footplate; it extends the full width of the footplate, and is made so low that a man can stand upright on the top of the tank and safely pass under any of the bridges. The exhaust steam dome is 11½ inches in diameter and 4 feet 6 inches high. The capacity of the tank is 1,600 gallons, and the coal space, which is made in horseshoe shape, is sufficient for 6½ tons of coal. The total length of the engine and tender over the buffers is 48 feet 6 inches. In order to heat the feed water in the tank, part of the exhaust steam from the cylinders is sent into it, passing through a set of small pipes running longitudinally in the bottom of the tender and finally up the steam dome into the tank. A stop valve is provided in the blast pipe, worked from the footplate, also another in the main exhaust pipe, which is 4 inches in diameter, in order to shut nearly all the blast off the fire when shunting at stations, so as to avoid waste of steam at the safety valves, and also to warm the water in the tender. By this means the feed-water is generally heated to boiling point, but still further to avoid waste a steam donkey pump is fixed on the footplate, the diameter of the cylinder being 6 inches and the length of stroke 6 inches. The pump would supply the boiler should the other pumps fail. The ordinary pumps for supplying the boiler are worked from the crosshead of the piston rod, having a 26-inch stroke. They each have two suction and two delivery clacks to facilitate the working, and an air-vessel is fixed over the top feed pipe. Turning to the motion, the piston rods are of steel, and are each made in one piece with its crosshead, the part of the rods which enters the pistons, being made with a sharp taper, so that on taking off the nut on the front side, the piston rod can be at once drawn through the stuffing-box of the back cylinder cover. The glands and packing rings are made in halves, being held together by a ring. The connecting rod is 6 feet 6 inches in length, and has big-ends of the marine pattern, the brasses being held by a cap secured by two bolts. The big-end brasses are lined with white metal. The valve motion is of the Stephenson shifting link type, having a coiled spring placed on the end of the weigh-bar shaft instead of balance weights. The motion is reversed by a lever, the rod being underneath; the cranks of the outside rods are placed on the same centres as the piston cranks. The weight of these engines

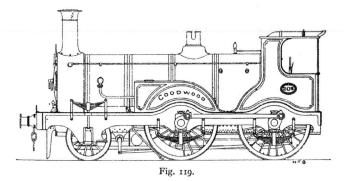

Fig. 119.

in working order is 38 tons 12 cwt, the leading wheels carrying 13 tons, the driving 14 tons 12 cwt, and the trailing 11 tons. In 1873 No. 85 was re-numbered 83, and in July, 1880, was again altered to 407; while in September, 1880, No. 84 was re-numbered 408. No. 408 is now fitted with the Westinghouse air brake. Both engines now have 18¼ inch cylinders, and the donkey feed pumps on the footplates have been removed. No. 408, however, has a duplex Westinghouse pump which will fill the boiler. These locomotives were built with copper chimney caps and painted a dark green, lined out with one broad band of black; but when the brake was added a fine red line was put round the black band, this being the standard practice. The number-plate is of brass, with a blue ground round the figures. The sandboxes are made in one with the leading splashers, and have valves operated from the footplate.

In November, 1872, two very fine four-wheels coupled express locomotives were built at Brighton Works from the design of Stroudley. They were numbered 201 and 202 and named *Belgravia* and *Goodwood* respectively, *(Fig. 119)*. The cylinders, which are horizontal, are 17 inches in diameter, and have a stroke of 24 inches, their centres being placed 2 feet 6 inches apart. The leading wheels are 4 feet in diameter and the driving and trailing 6 feet 6 inches. The overhang at the leading end from the front of the buffer beam is 4 feet 8½ inches and at the trailing end 4 feet 2 inches. Wheelbase is 16 feet 8 inches, being 8 feet 4 inches between each pair of centres. The boiler is telescopic and made after Stroudley's usual practice, 4 feet 3 inches in diameter outside the largest ring, the length of the barrel being 10 feet 5 inches and the height of the centre line above the rail level 7 feet. The boiler contains 160 tubes 1½ inches in diameter, with a length of 10 feet 11¾ inches. The length of the firebox casing outside is 6 feet 2 3/16 inches, the width 4 feet 0½ inch and the depth 4 feet 11 inches both at the front and back. The inside firebox roof was formerly circular, and connected to the outer casing by rigid iron stay bolts, but now the ordinary flat-topped firebox is used and the standard girder roof stay bars. When new these engines had no domes, but had safety valves like those of the goods engines Nos 84 and 85; these, however, were all done away with about the same time. The heating surface of the tubes is 1,172 sq, ft. and that of the firebox 112, making a total of 1,284 sq. ft. The weight in working order is 41 tons 4 cwt, distributed as follows: 13 tons on the leading wheels, 15 tons on the driving wheels and 13 tons 4 cwt on the trailing wheels. As will be seen from the diagram, these engines have outside framings, the driving wheels having inside bearings as well as outside. Just previously to the death of Stroudley the leading wheels were fitted with inside bearings, so as to relieve the weight on the outside boxes. No horn blocks were used for the inside boxes to work in, the latter being loose on the journals. The motion of these engines is of the Stephenson type, having the weigh-bar shaft below and reversed by a lever. The motion-plate is brought high up above the framing, allowing the barrel of the boiler to take a bearing on it. These engines were eventually fitted with the Westinghouse brake. No date plates were ever put on them, for some reason or other, and the general impression is that they were converted from Stephenson's single engines, but such is not the case. When new these locos ran with tenders of Craven's design, but about 1876 they were fitted with Stroudley's large express tenders having outside bearings. The tender wheels are 4 feet in diameter, with the bearing springs underhung. The wheelbase is 12 feet, being equally divided. The overhang at the leading end is 4 feet 3¼ inches and 3 feet 9¼ inches at the trailing end. The capacity of the water tank, which has a well, is 2,520 gallons. The weight of the tender, exclusive of coal, of which about 6½ tons can be carried, is 26 tons 5 cwt, 7 tons 13 cwt being on the leading wheels, 9 tons 15 cwt on the middle and 8 tons 17 cwt on the trailing. Half of the exhaust steam is discharged through twelve copper pipes and up the exhaust dome into the tender tank to heat the feed water, the boiler being fed by two long-stroke pumps. These locomotives in January, 1897, were re-numbered 501 and 502 respectively. No. 501 was scrapped in January, 1899. No. 502 was re-numbered 602 in June, 1900, and scrapped December, 1901.

The next class of locomotive which Stroudley introduced were the small six-wheels coupled tanks, better known as the Terriers, owing to their small and smart appearance, coupled with their ability to get quickly away with a train. These were specially designed for working the East London Railway between New Cross and Liverpool Street and the South London line between Victoria and London Bridge, the traffic on which at the time was very small and did not pay working expenses. The permanent way also was light and in bad order, the rails being made of iron. Very severe gradients are found on both routes, though for short distances. Under these conditions the engines had to be made as light as possible, but their success from the first was remarkable. The first came out in October, 1872, and the last in September, 1880, their total number being 50. Illustrated by *Figs 120 and 121*, the former diagram showing them as they came

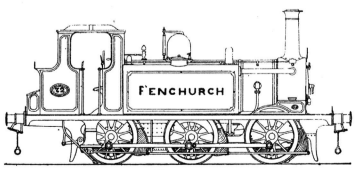

Fig. 120.

out new and the latter as they are at present. The cylinders were originally 13-inch diameter and the stroke of the piston 20 inches, but some now have 14-inch cylinders. The centres of the cylinders are 2 feet 3 inches apart. They are placed at an angle of 1 in 11. The travel of the slide valves is 3⅝ inches. The height of the blast pipe above the boiler centre line is 11 inches, the diameter of the orifice being 3½ inches. From the centre of the cylinders to the centre of the crank axle is 7 feet 8 inches. The diameter of the coupled wheels is 4 feet. From the front of the buffer beam to the leading centre is 4 feet 7¼ inches and the overhang at the trailing end is 6 feet 7¼ inches. Wheelbase is 12 feet equally divided. The boiler, which is made in three rings of 7/16 inch iron plate, is 3 feet 5⅛ inches in diameter outside the middle ring, the length of the barrel being 7 feet 10 inches and the height of the centre line above the rail level only 5 feet 8⅝ inches. The length of the firebox casing outside is 4 feet 1 inch and the breadth 3 feet 6⅞ inches, the depth below the centre line of the boiler being 3 feet 10½ inches, both at the front and at the back. The inside firebox roof is inclined from the front tube plate to the back plate of the box, and the outer shell is stayed to the inner box by rigid roof stays. Several of these engines some years ago had steel fireboxes, which although they appeared to give satisfaction for a considerable time were afterwards superseded by copper boxes. These engines have 121 tubes 1¾ inches in diameter and 8 feet 4¼ inches long. They were inclined from the front tube plate to the back. The height of the chimney above the rail is only 11 feet 3 inches. The heating surface of the tubes is 463 sq. ft and that of the firebox 55 sq. ft, giving a total of 518 sq. ft. The grate area is 10 sq. ft. The boiler is supplied by two pumps, worked by the crossheads, and carries a working pressure of 140 lbs per sq. in. The capacity of the tanks is 500 gallons and the coal bunker has 27 cubic feet of coal space. Total length of these little engines over the buffers is 26 feet 0½ inch. The mean weight in working order is 24 tons 7 cwt, which is distributed as follows: 8 tons 5 cwt on the leading wheels, 8 tons 2 cwt on the driving wheels and 8 tons on the trailers. The motion is of the Stephenson type, having the weigh-bar shaft above, and is reversed by a lever. The regulator is of the standard pattern. In 1878 No. 40, *Brighton*, was sent to the Paris Exhibition, where it was awarded a gold medal. On its return from Paris to Dieppe after the Exhibition it worked a train in a manner that gave great satisfaction to both English and French authorities. When this class was first introduced a copper connecting pipe went across

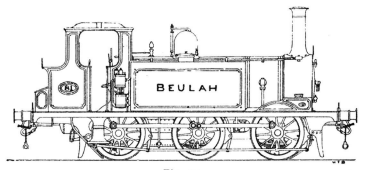

Fig. 121.

the boiler to allow the exhaust steam to circulate, but the water also got through, the result being that the pipes were removed. A hand brake only was fitted on when new, but all were eventually equipped with the Westinghouse brake. When No. 40 went to Paris it was fitted with Stroudley's patent speed indicator, but was the only engine of its class to have one; this was subsequently taken off. Until quite recently No. 40 had 'Gold Medal, Paris Exhibition, 1878' painted in gilt letters on the tank sides, but this has now been dispensed with. It may also prove of interest to state that a negro used to fire on this engine. So successful were these engines that some are still working the trains usually taken by Stroudley's larger tanks, coming round the South London line morning and evening with ten and twelve closely-packed coaches, and it must be borne in mind that most of the stations are less than a mile apart.

Here are numbers, names and dates of the class, from which it will be seen that most of the engines were built backwards – the highest number first, down to the lowest:

No.	Name	Date	No.	Name	Date
35	Morden	June, 1878.	60	Ewell	"
36	Bramley	"	61	Sutton	"
37	Southdown	May "	62	Martello	"
38	Millwall	"	63	Preston	"
39	Denmark	May, 1878.	64	Kemptown	June, 1874
40	Brighton	Mar "	65	Tooting	"
41	Piccadilly	June, 1877	66	Hatcham	"
42	Tulse Hill	"	67	Brixton	"
43	Gipsy Hill	"	68	Clapham	August, 1874
44	Fulham	"	69	Peckham	"
45	Merton	"	70	Poplar	Nov, 1872.
46	Newington	Dec, 1876.	71	Wapping	Oct "
47	Cheapside	"	72	Fenchurch	Nov "
48	Leadenhall	"	73	Deptford	"
49	Bishopsgate	"	74	Shadwell	"
50	Whitechapel	"	75	Blackwall	"
51	Rotherhithe	"	76	Hailsham	June, 1877
52	Surrey	Dec, 1875.	77	Wonersh	June, 1880
53	Ashtead	"	78	Knowle	"
54	Waddon	"	79	Minories	"
55	Stepney	"	80	Bookham	"
56	Shoreditch	"	81	Beulah	July
57	Thames	"	82	Boxhill	"
58	Wandle	Nov "	83	Earlswood	Aug
59	Cheam	Oct "	84	Crowborough	Sept

These locomotives are painted the standard 'Stroudley's improved engine green' and lined out with green and black bands picked out with red and white. The angle-iron round the foot-plating is crimson, having a black band to the edge, with a red and orange fine line on either side. No. 72, *Fenchurch*, was sold in June, 1898, to the Newhaven Harbour Company for working the ballast wagons on the new dock and wharf. Its number was, of course, removed, as well as the Westinghouse brake and the electric passenger and guard communication. Engines Nos 77–84 have cast-iron brake blocks, all the others having wooden ones. No. 84 was the only one of the class built with 14-inch cylinders, but Nos 35, 36, 38–40, 43, 45, 46, 48–50, 53–55, 57, 60, 62–

64, 66, 74, 75, 78, 81 and 83 were afterwards fitted with cylinders of that size. When the 14-inch cylinders were put in the whole of the exhaust from the cylinders and air brake pump was turned up the chimney, the boiler being fed with cold water.

The following engines of this class have now been sold or scrapped:

No.		Their No.
40	Sold to the Isle of Wight Central Railway, Jan 1902	11
41	Scrapped in March, 1902	
46	Sold to the LSWR, March, 1903	734
48	Scrapped in July, 1901	
51	”	
58	” in February, 1902	
60	” in December, 1902	
65	” in February, 1901	
66	”	
68	Sold to the LSWR, March, 1903	735
69	Sold to the Isle of Wight Central Railway, April, 1900	10
70	Rother Valley Railway, May, 1901	3 *Bodiam.*
72	Newhaven Harbour Company	
74	Isle of Wight Central Railway, March, 1899	9
75	”	

Excepting engines sold or scrapped, Nos 42–74 have been re-numbered 642–674.

To cope with the increasing number and weight of the goods trains, Stroudley built some more goods engines similar to Nos 83 and 84 (Class 'C'), which he had decided to adopt as his standard. The first engine of this second lot was numbered 77, and commenced to run in March, 1873. In a few details only, the first two and the later ones differed, the chief being the addition of a dome. The donkey pump was abandoned and the tender made a few inches higher, the capacity being increased from 1,600 gallons to 1,860. The tender frames differed slightly, which will be noticed better by comparing the diagrams, and the exhaust dome was not so high. The framings of the engines also slightly differed and the apparatus for controlling the action of the blast was discarded. The oil boxes, instead of being on the top of the driving splashers, were placed behind. In 1873 Kitson & Co., of Airedale Foundry, Leeds, accepted the contract for twelve engines of this class. As previously stated, the cylinders of these engines were 17½ inches in diameter, but nearly all have now been fitted with 18¼ inch cylinders. The engines built by Kitson & Co. were numbered 85 to 96, *(Fig. 122)*. The following list

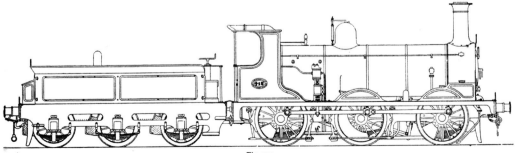

Fig. 122.

shows the numbers, makers, makers' numbers and dates of this class, including the first two, as well as the numbers to which they afterwards were altered:

Engine No.	Makers	Maker's No.	Date New	No. altered to	Date Renumbered
77	LBSCR	–	March, 1873	401	June 1880
78	LBSCR	–	April 1873	402	June 1880
79	LBSCR	–	April 1873	403	June 1880
80	LBSCR	–	June 1873	404	June 1880
81	LBSCR	–	June 1873	405	June 1880
82	LBSCR	–	June 1873	406	June 1880
83	LBSCR	–	Dec 1871	407	July 1880
84	LBSCR	–	Dec 1871	408	Sept 1880
85	Kitson & Co.	1877	June 1873	409	Oct 1882
86	Kitson & Co.	1878	July 1873	410	Feb 1883
87	Kitson & Co.	1879	Sept 1873		Feb 1883
88	Kitson & Co.	1880	Oct 1873	412	Feb 1883
89	Kitson & Co.	1881	Occ 1873	413	Feb 1883
90	Kitson & Co.	1882	Dec 1873		Feb 1883
91	Kitson & Co.	1883	March 1874	415	Sept. "
92	Kitson & Co.	1884	April 1874	416	Feb 1883
93	Kitson & Co.	1885	June 1874	417	Feb 1883
94	Kitson & Co.	1886	Sept. 1874	418	Feb 1883
95	Kitson & Co.	1887	Nov 1874	419	Feb 1883
96	Kitson & Co.	1888	Nov 1874	420	Feb 1883

The date plates of these engines are on the back of the tenders, and not on the engines. Nos 408, 409, 410, 411, 413, 415, 419 and 420 are fitted with the Westinghouse brake and electrical passenger and guard communication, so as to enable them to work excursion trains at holiday times. When the goods tender engines had the air brake they were lined out with a fine red line on either side of the broad black band round the cab, splashers, tanks and boiler. An exception to this rule was made with engine No. 418, which had the red lining, but no brake, although studs for supporting the brake-block hangers were on the framings. Latterly this engine was painted like the others without the brake.

In March, 1873, a small tank engine was put on the West London Extension line from Clapham Junction to Kensington. It was built by Sharp, Stewart & Co., of the Atlas Works, Manchester, in August, 1872, and numbered in their books 2242. Craven, the late locomotive superintendent, had an interest in the Tunis Railway and also in a line in Jersey. He ordered a small tank engine from Sharp for the Tunis line, but as it was not ready to go out by the time desired Craven arranged for an engine that was also under construction for the Jersey Co. to go in its stead. When the other engine was complete and ready for delivery the Jersey Co. refused it, so it was thrown on Craven's hands, as he was responsible for it. He, however, used his influence with the directors of the LBSCR, who agreed to purchase it, as a small engine was wanted to pull three coaches on the West London Extension Railway. As the traffic increased and more coaches were required on this service, the engine was found not powerful enough for the work, and was replaced by one of Stroudley's 'Terrier' or 'A' class engines. This engine was then transferred to Newhaven, where it shunted the ballast

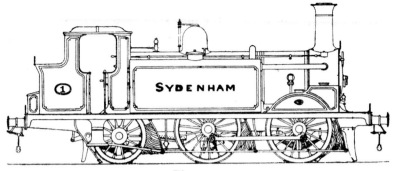

Fig. 123.

trucks on the new harbour and breakwater works. While it was down there it was named *Bishopstone*. It was afterwards transferred to Havant, where it worked the South Hayling line, and was then named *Fratton*. This engine was originally numbered 53; in November, 1875, it was re-numbered 270, in April, 1880, 357, and in April, 1886, 497; it was sold in December, 1890. The cylinders, which were inside, were 12 inch diameter, with a stroke of 17 inches. The driving and trailing wheels were 4 feet in diameter and the leading wheels 2 feet 9 inches. The wheel-base was 12 feet 5 inches and the framings were inside. The boiler was 9 feet long and 3 feet 3 inches in diameter, containing 108 tubes of 2 inch diameter. Two spring balance safety valves were placed over the firebox. This engine had a very large dome and a polished copper chimney cap. The boiler was fed by injectors and had a raised firebox. The heating surface was 570 sq. ft, of which the tubes gave 523 sq. ft and the firebox 47 sq. ft. When new this engine had only a front weather board, but Stroudley carried it right over to the back. It was originally printed bright green and had a polished dome and safety valve casing. The side tanks had a water capacity of 400 gallons.

On 25 November, 1873, the first of Stroudley's celebrated 'D' class passenger tank engines, No. 1, *Sydenham*, commenced work. This type of engine was designed and constructed to meet the requirements of the ever-increasing London suburban traffic of the LBSCR. They were very up-to-date for their time and continued to be built down to 1887. At that time they were often employed in working goods trains and through trains to Brighton *via* Oxted and Lewes. The fact of there being 125 of this class is sufficient to prove their efficiency. They are illustrated by *Figs 123 and 124*, the latter showing them as reboilered by Billinton, the present chief mechanical engineer. The engines are numbered 1–34, 221–299, and 351–362. The cylinders are 17 inches in diameter, placed at an angle of 1 in 11½, and 2 feet 2 inches between centres. The stroke of the pistons is 24 inches. The steam ports are 1½ inches wide and the exhaust ports 2½ inches, arranged in two groups each 7½ inches long, the exhaust steam from the lower port being carried completely round the outside of the cylinders. The total wheelbase is 15 feet, being 7 feet 7 inches between leading and driving centres and 7 feet 5 inches between driving and trailing. The overhang of the frames at the leading end is 5 feet 11¼ inches and at the trailing end 7 feet 10¼ inches. When new, with 3 inch tyres, the coupled wheels are 5 feet 6 inches in diameter and the trailing wheels

4 feet 6 inches. The boilers are made in three rings, the external diameter of the largest being 4 feet and the length of the barrel 10 feet 2 inches. This latter is the standard dimension for all classes of Stroudley engines except the 'A' or 'Terrier' class. The height of the centre line above the rail level is 6 feet 11 inches. The length of the firebox casing is 5 feet 2¼ inches, the breadth 4 feet 1 inch and the depth below the boiler centre line 5 feet at the front and 4 feet 6 inches at the back. The inside box had originally a semi-circular roof stayed direct to the outside casing by rigid stays. Now, however, all the engines have a flat-topped firebox and girder roof stay bars. Originally these engines had 175 brass tubes 1¾ inch diameter and 10 feet 8¾ inches long, the heating surface being – tubes 952 sq. ft, firebox 91 sq. ft, total 1,043 sq. ft. The grate area is 15 sq. ft and the working pressure 150 lbs per sq. in. The capacity of the water tanks is 860 gallons and of the coal bunker 30 cwt. The total length over the buffers is 31 feet 7½ inches and the height of the chimney above the rail level 13 feet. All these engines are fitted with the Westinghouse air brake and the electrical communication between the passengers, guards and drivers, and originally all were fitted with Stroudley's speed indicator. No. 25, *Rotherfield*, was the first engine upon this railway to be fitted with the Westinghouse air brake. No. 27, *Uckfield*, was some years ago fired with petroleum, on Tarbutt's system, but it was not very successful, the burning of liquid fuel not having been brought to perfection at that time. No. 297, *Bonchurch*, was in the fatal accident at Mayfield on 1 September, 1897. No. 4, *Mickleham*, is noteworthy for having run 62,000 miles without any repairs. Several engines of this class have been reboiled by Billinton. The dimensions of the barrel are identical with the old ones, but in some cases, especially the first engines reboiled by Billinton, the steam dome is placed on the middle ring and a manhole over the firebox. Latterly these new boilers have the dome in the original position on the back ring and no manhole over the firebox. The grate area is 15.3 sq. ft and the boiler contains 207 tubes 1⅝ inch diameter, giving a heating surface of 924.57 sq. ft, while that of the firebox is 82.06 sq. ft, making a total of 1,006.63 sq. ft. When these engines are reboiled all the exhaust steam from the cylinders is turned up the chimney instead of half going into the tanks to heat the feed water, the pumps being removed and Gresham & Craven's combination injectors fitted instead on the back plate of the firebox. The total weight of these engines in working order is 38 tons 10 cwt, – 13 tons 10 cwt being on the leaders, 13 tons 10 cwt on the drivers and 11 tons 10 cwt on the trailing wheels. Nos 1–34 and 268–299 have wooden

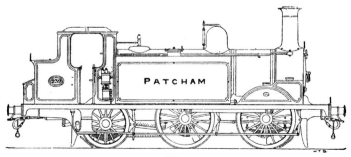

Fig. 124.

brake blocks, whilst Nos 221–267 and 351–362 have iron ones. Nos 235–267 were built by Neilson & Co., Hyde Park Locomotive Works, Glasgow, the makers' numbers of 234–267 being 2703–2736 and of 233, 2938. The reason for No. 233 coming last is because it was kept as a pattern engine, a large portion of its standard details having been made at Brighton Works. All the other engines of the class were built at Brighton, and the following is a complete list of the numbers, names and dates when new:

No.	Name	Date	No.	Name	Date
1	Sydenham	Nov., 1873	230	Brookhouse	Oct., 1884
2	Wandsworth	Dec., 1873	231	Horsham	June, 1884
3	Battersea	"	232	Lewes	"
4	Mickleham	"	233	Handcross	March, 1883
5	Streatham	"	234	Rottingdean	Oct., 1881
6	Wimbledon	"	235	Broadwater	Nov, 1881
7	Bermondsey	March, 1874	236	Ardingly	"
8	Brockley	March, 1874	237	Cuckfield	"
9	Anerley	"	238	Lindfield	"
10	Banstead	"	239	Patcham	"
11	Selhurst	"	240	Ditchling	"
12	Wallington	June, 1874	241	Stanmer	"
13	Pimlico	Dec., 1874	242	Ringmer	Nov., 1881
14	Chelsea	"	243	Ovingdean	"
15	Brompton	"	244	Hassocks	"
16	Silverdale	March, 1875	245	Withdean	"
17	Dulwich	"	246	Bramber	"
18	Stockwell	May, 1875	247	Arlington	"
19	Belmont	June, 1875	248	Ashurst	"
20	Carshalton	"	249	Hilsea	"
No.	Name	Date	250	Hoatbly	"
21	Beddington	"	251	Singleton	"
22	Addington	"	252	Bnckhurst	"
23	Mayfield	Aug., 1875	253	Pelham	"
24	Brambletye	Oct., 1875	254	Hambledon	Nov., 1882.
25	Rotherfield	March, 1876	255	Willingdon	"
26	Hartfield	"	256	Stanford	"
27	Uckfield	April, 1876	257	Brading	"
28	Isfield	"	258	Cosham	"
29	Lambeth	"	259	Telford	"
30	Camberwell	"	260	Lavington	"
31	Borough	May, 1876	261	Wigmore	"
32	Walworth	"	262	Oxted	"
33	Mitcham	"	263	Purley	"
34	Balham	June, 1876	264	Langston	"
35	Southwark	"	265	Chipstead	"
36	New Cross	"	266	Charlwood	"
221	Warbleton	June, 1885	267	Maresfield	"
222	Cuckmere	"	268	Baynards	June, 1880.
223	Balcombe	"	269	Crawley	May, 1880.
224	Crowhurst	May, 1885	270	Warnham	"
225	Ashburne	"	271	Eridge	"
226	Westham	"	272	Nevill	"
227	Heathfield	Dec., 1884	273	Dornden	"
228	Seaford	"	274	Guildford	Dec. 1879
229	Dorking	No., 1884	No.		

Name	Date		No.	Name	Date
275	Cranleigh	,,	293	Norbury	,,
276	Rudgwick	,,	294	Rosebery	,,
277	Slinfold	,,	295	Whippingham	Oct., 1877
278	Groombridge	,,	296	Osborne	,,
279	Tunbridge Wells	,,	297	Bonchurch	,,
280	Grinstead	Oct., 1879	351	Chailey	Dec., 1885
281	Withyham	,,	352	Lavant	,,
282	Rowfant	,,	353	Keymer	,,
283	Aldgate	,,	354	Lancing	April, 1886
284	Ashburnham	,,	355	Worthing	May, 1886
285	Holmwood	,,	356	Coulsdon	June, 1886
286	Ranmore	June, 1879	357	Riddlesdown	,,
287	Buryhill	,,	358	Henfield	Oct., 1886
288	Effingham	,,	359	Egmont	Dec., 1886
289	Holmbury	,,	360	Leconfield	,,
290	Denbies	,,	361	Upperton	,,
291	Deepdene	,,	362	Kidbrooke	March, 1887
292	Leigham	Nov., 1877			

All this class are still at work, but in May, 1878, Nos 35 and 36 were re-numbered 298 and 299 respectively. In March, 1897, No. 294 *Rosebery* became *Falmer*; in July, 1897, No. 272, *Nevill*, was re-named *Goring*; and in April, 1898, No. 259, *Telford*, was re-named *Barnham*. In December, 1901, No. 296 *Osborne* was re-named *Peckham*.

In December, 1874, Stroudley built his first single-wheel express engine, the famous *Grosvenor*, No. 151. This locomotive was originally styled class 'B', but is now known as class 'G'; *Fig. 125*, it has cylinders of 17 inch diameter by 24 inch stroke, placed horizontally with the centres 2 feet 2 inches apart. The length of the steam ports is 1 foot 3 inches and the breadth 1½ inches, the exhaust ports being 1 foot 3 inches by 2 inches. The travel of the valves is 3¾ inches, this being the standard for all Stroudley's engines, the lead ⅛ inch, and the outside lap ⅞ inch. The diameter of the blast pipe orifice is 4¾ inches and the height above the centre line of the boiler 1 foot 0¼ inches. The leading and trailing wheels are 4 feet 6 inches in diameter and the driving 6 feet 9 inches. The overhang at the leading end is 5 feet 8¼ inches and at the trailing end 4 feet 3¼ inches. The wheelbase is 15 feet 9 inches, from leading to driving centres being 8 feet and from driving to trailing 7 feet 9 inches. The boiler barrel is made in three rings, the plates being ½ inch thick; its largest external diameter is 4 feet 5 inches, from which it will be seen that Stroudley was well ahead of the times as to passenger engine designing; the length of the barrel is 10 feet 5 inches. The length of the firebox outside is 6 feet 2¼ inches and the breadth 4 feet 1 inch, the covering plate being 9/16 inch thick. The height of the inside box from the firebars to the crown is 5 feet 11 inches. The boiler originally had 220 tubes of 1¾ inch diameter and 2 of 1½ inches, but these are now superseded by 307 tubes of 1⅝ inches in diameter. The heating surface of the tubes is 1,100 sq. ft, and that of the firebox 110 sq. ft, giving a total of 1,210 sq. ft. The grate area is 19.3 sq. ft. The centre line of the boiler is 7 feet 3½ inches, and the height of the chimney 13 feet 2 inches above the rail level. From the centre of the cylinders to the centre of the crank axle is 9 feet 9⅝ inches. The frames, which are inside, are 1 inch thick. The clack-boxes are on the side of the boiler, and the working pressure is 150 lbs per sq. inch. The inside firebox was originally connected to the outer shell by rigid stays, but these are now replaced by girder roof stay-bars. The weight of the *Grosvenor* full is 33 tons, distributed as follows: 10 tons 8 cwt on the leading wheels, 14 tons on the driving, and 8 tons 12 cwt on the trailing wheels. The engine is fitted with two pumps worked direct from the crossheads; these are specially designed to pump hot water at high speed. The *Grosvenor* was originally fitted with a powerful steam brake, and also a hand-brake on the engine footplate, the latter being a very unusual arrangement. Wooden brake-blocks acted on the front of the driving and trailing wheels, but these have now been replaced by iron blocks on the driving wheels only. There are two blocks to each wheel, arranged similarly to those on the 'Stephenson' class, which will be shown in due course, whilst the steam brake has been replaced by the Westinghouse. The *Grosvenor* went, in 1875, to the Newark brake trials to work a train fitted with a Westinghouse vacuum brake, but the compressed air system proving more satisfactory, it was adopted on the LBSCR. On 13 August, 1875, No. 151 was the first engine to make the through run from Victoria to Portsmouth, the distance of 87 miles being run in 1 hour 50 mins. This engine is the only one of its class, and in January, 1880, was re-numbered 326. When the *Grosvenor* was built it was fitted with one of Craven's old tenders, but this was replaced by one of Stroudley's large outside framed tenders to go to Newark with. The latter has now been removed and a standard inside framed tender substituted. The driving springs of this engine were originally of the volute type, but are now spiral. The leading and trailing springs are of the ordinary laminated pattern.

In 1875, Stroudley designed two four-coupled express engines, numbered and named 206 *Carisbrooke*, and 207 *Freshwater*. Built at the Brighton Works in November and December of that year respectively, they were originally called class 'B', their boilers being very similar to that of No. 151. They were built for the 8.45 a.m. Brighton to London Bridge express, which they worked until 1880. The cylinders, which were placed horizontally, were 17 inch diameter, the stroke of the pistons being 24 inches. The leading wheels were 4 foot 1 inch diameter and the driving and trailing 6 feet 6 inches. The wheel-base was 16 feet 8 inches, equally divided. The boiler was made in three rings, the largest external diameter being 4 feet 5 inches, the length 10 feet 5 inches, centre line from rail level 7 feet 1 7/16 inches. It contained 206 tubes 1¾ inch

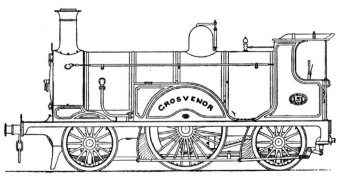

Fig. 125.

diameter, 10 feet 8¾ inches long. The length of the firebox casing is 6 feet 2¼ inches, the depth below the centre line 5 feet 6 inches at the front and 4 feet 1 inch at the back. The heating surface of the tubes is 1,022 sq. ft and of the firebox 110 sq. ft, giving a total of 1,132 sq. ft. These two engines were very similar to the Belgravia class *(Fig. 119)* in external appearance, but they always had steam domes, and were built new with the Westinghouse brake. They had outside frames, the leading and driving wheels having both inside and outside bearings. The valve motion was of the Stephenson type, having the weigh-bar shaft below. All the working parts were made so as to be interchangeable with Nos 201, 202, 204 and 205. The weight of Nos 206 and 207 was 39 tons 3 cwt, the leading wheels carried 13 tons 1 cwt, the driving 15 tons 1 cwt, and the trailing 11 tons 1 cwt. They had old tenders originally belonging to Craven's engines. In March, 1897, No. 206 was re-numbered 506, and in May, 1899, was again renumbered 606; in June, 1897, 207 was re-numbered 507, and 607 in December, 1900.

In November, 1874, Stroudley built his first goods tank engine, No. 97 *Honfleur*, class 'E'. This class, of which there are no less than seventy-two engines, the numbers running from 85–156, *(Figs. 126 and 127)*, the former showing them in their original condition and the latter as rebuilt. The cylinders are 17 inches in diameter and the stroke of the pistons is 24 inches, the centres being placed 2 feet 2 inches apart at an inclination of 1 in 11½. They have the same sized steam ports, and are of the same design throughout as the 'D' class. The overhang at the leading end is 6 feet 0½ inch and at the trailing 8 feet 1¼ inches. The wheelbase is 15 feet 3 inches, being 7 feet 6 inches between the leading and driving centres and 7 feet 9 inches between the driving

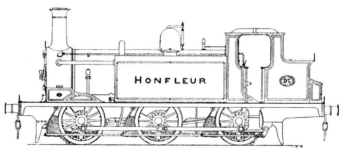

Fig. 126.

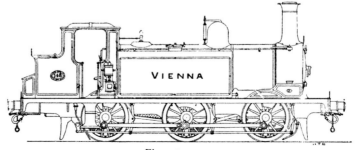

Fig. 127.

and trailing. The diameter of the wheels when new, with 3 inch tyres, is 4 feet 6 inches. The mean diameter of the boiler outside is 3 feet 11 inches, the length 10 feet 2 inches, and the height of the centre line above the rail level 6 feet 11 inches. The length of the firebox casing is 5 feet 2¼ inches, the depth below the centre line of the boiler being at the front 4 feet 9 inches, and at the back 4 feet 3 inches. Originally these engines had 160 tubes 1¾ inch diameter and 10 feet 8¾ inches long, the heating surface being 772 sq. ft, and that of the firebox 80 sq. ft, giving a total of 852 sq. ft. They now have 231 tubes 1½ inch diameter, the heating surface of which is 952 sq. ft. The internal firebox originally had a circular roof stayed directly to the crown, but it now has a flat roof with girder stays; this has increased its heating surface by 7 sq. ft, bringing the total up to 1,039 sq. ft. The total length over the buffers is 32 feet 4½ inches. The capacity of the two side tanks is 900 gallons, and that of the coal bunker 1 ton 15 cwt. Several of these engines have been rebuilt by Billinton, the dimensions of their new boilers being identical with the old ones, except that they have 207 tubes 1⅝ inch diameter with a heating surface of 924.57 sq. ft, whilst the firebox has 82.06 sq. ft, making a total of 1,006.63 sq. ft. It will be seen from the diagram that the dome is placed on the second ring, and there are a few minor alterations. As in the 'D' class, the exhaust steam does not go into the tanks. The weight of these locomotives in working order is 39 tons 10 cwt, the leading wheels carrying 13 tons 10 cwt, the driving 13 tons 10 cwt, and the trailing 12 tons 10 cwt. The following engines of this class are fitted with the Westinghouse air brake: Nos 85–103, 106–111, 113–115, 117, 119, 120, 122, 123, 125–132, 137–141. Nos 145–156 were built new with the Westinghouse brake, and these have iron brake blocks, whereas the remainder have wooden ones. Those that are not fitted with the Westinghouse brake have a powerful steam brake. Originally all goods tank engines fitted with the air brake were painted the standard passenger engine colour, the following being so painted: Nos 85–96, 117–119, 123, 128, 129 and 145–156. All the engines of this class were built at Brighton, and the following is a complete list of their numbers, names and dates when new:

No.	Name	Date	No.	Name	Date
85	Cannes	Feb., 1883	103	Normandy	Oct., 1876
86	Geneva	March, 1883	104	Brittany	"
87	Bologna	"	No.	Name	Date
88	Rhine	"	105	Morlaix	"
89	Brest	April, 1883	106	Guernsey	"
90	Berne	"	107	Alderney	Nov., 1876
91	Fishbourne	Oct., 1883	108	Jersey	"
92	Polesden	"	109	Strasbourg	March, 1877
93	Calbourne	"	110	Burgundy	"
94	Shorwell	"	111	Montpellier	"
95	Luccombe	Nov., 1883	112	Versailles	April, 1877
96	Salzberg	"	113	Granville	"
97	Honfleur	Nov., 1874	114	Trouville	May, 1877
98	Marseilles	"	115	Lorraine	June, 1877
99	Bordeaux	Dec., 1874	116	Touraine	"
100	Calvados	April, 1875	117	Florence	Aug., 1877
101	Orleans	"	118	Trocadero	Sept., 1877
102	Cherbourg	"	119	Rochelle	"

No.	Name	Date	No.	Name	Date
120	Provence	"	140	Toulouse	"
121	Verona	July, 1878	141	Mentone	"
122	Leghorn	Aug., 1878	142	Toulon	"
123	Seine	"	143	Nuremberg	"
No.	Name	Date	144	Chambery	April, 1879
124	Bayonne	"	145	France	Oct., 1880
125	Navarre	"	146	Havre	"
126	Gascony	"	147	Danube	"
127	Poitiers	Oct., 1878	148	Vienna	"
128	Avignon	"	149	Lucerne	"
129	Alençon	"	150	Adriatic	Nov., 1880
130	Rennes	"	151	Helvetia	Dec., 1880
131	Gournay	"	152	Hungary	"
132	Epernay	"	153	Austria	March, 1881
133	Picardy	"	154	Madrid	"
134	Ancona	Dec., 1878	155	Brenner	"
135	Foligno	"	156	Munich	"
136	Brindisi	"			
137	Dijon	"			
138	Macon	"			
139	Lombardy	March, 1879			

In 1883, No. 113, *Granville*, was re-named *Durdans*.

In January, 1877, the *Abergavenny*, No. 325, was put to work. This fine locomotive, formerly class 'F', but now class 'G', is illustrated by *Fig. 128*. The cylinders are small, being only 16 inches in diameter, with a stroke of 22 inches. They are horizontal, with the centres 2 feet 2 inches apart. The leading and trailing wheels are 4 feet 6 inches in diameter and the driving 6 feet 6 inches. The wheelbase is 15 feet 9 inches, the leading and driving centres being 8 feet apart and the driving and trailing 7 feet 9 inches. The overhang at the leading end is 5 feet 8¼ inches and at the trailing end 3 feet 5¼ inches. The boiler barrel is 4 feet 3 inches in diameter outside the largest ring and 10 feet 2 inches long, its centre line being 7 feet 1½ inches above the rail level. The firebox casing is 5 feet, 8¼ inches long and its depth below the boiler centre line is 5 feet 2 inches at the front and 4 feet 8 inches at the back. The crown of the casing is also 1½ inches lower at the back than at the front, being parallel with the crown of the inside box, which has a similar drop. The boiler originally contained 202 tubes of 1¾

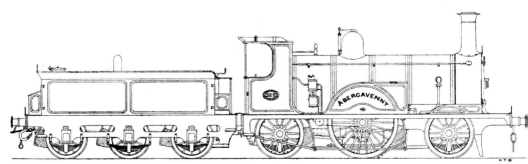

Fig. 128.

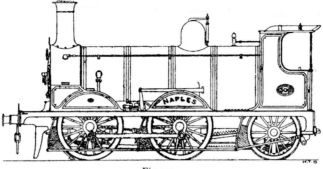

Fig. 129.

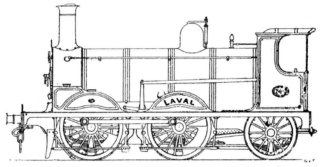

Fig. 130.

inch diameter, having a heating surface of 971 sq. ft, that of the firebox being 103 sq. ft and the total 1,074 sq. ft. It now has 244 tubes of 1½ inch diameter, with a heating surface of 1,006 sq. ft. The firebox crown and roof stay bars have been altered as in previously mentioned engines. A novel feature of the *Abergavenny* is that the wheels, which were made at the Brighton Works, when new were finished bright. The clack boxes and delivery pipes are on the side of the boiler. The brake blocks are of wood, acting on the front of the driving and trailing wheels. When built *Abergavenny* had Stroudley's large type of standard outside-framed tender, but now has one with inside-frames, the original tender being used with others on the goods engines. It used to be stationed at New Cross, and ran the night boat train to Newhaven for several years. Considering the smallness of the cylinders, it is still capable of some very good work, and is now stationed at Tunbridge Wells. The weight in working order is 32 tons 13 cwt, the leading wheels carrying 10 tons 9 cwt, the driving 14 tons 9 cwt and the trailing 7 tons 15 cwt.

In September, 1876, Stroudley introduced another type of locomotive, which was specially designed for working the fast fruit traffic between Worthing and London and fast fish and Grande Vitesse trains between Newhaven and London. These engines gave such satisfaction on those services that they were also used on excursion and heavy stopping passenger trains. They are classed as 'D 2', *(Figs 129 and 130)*. As will be seen, they have the leading and driving wheels coupled, the diameter being

5 feet 6 inches and that of the trailing wheels 4 feet 6 inches. The cylinders are of the standard design, 17 inch diameter by 24 inch stroke, with an inclination of 1 in 11½ to the centre line of the crank. The overhang at the leading end is 5 feet 11¼ inches and at the trailing end 3 feet 5¼ inches, the wheelbase being 15 feet 6 inches, 7 feet 7 inches of which is between the coupled centres and 7 feet 11 inches between the driving and trailing. The boiler is 4 foot 3 inch diameter outside the largest ring and 10 feet 2 inches long, its centre line being 7 feet 1 9/16 inches above the rail level. The length of the firebox casing is 5 feet 8¼ inches, the breadth 4 feet 1 inch and the depth below the boiler centre line 5 feet 2 inches at the front and 4 feet 1 inch at the back. As with previous engines, the rigid firebox stays have been replaced by those of the solid girder type, with a flat-topped inside box. There were formerly 202 tubes of 1¾ inch diameter, their heating surface being 971 sq. ft, whilst that of the firebox is 103 sq. ft, giving a total of 1,074 sq. ft, but now there are 244 tubes of 1½ inch diameter, having a heating surface of 1,006 sq. ft; the grate area is 17 sq. ft. The clack boxes and delivery pipes are on the side of the boiler, and all the engines are fitted with the Westinghouse brake, the brake blocks being of wood. They were all built at the Brighton Works, their numbers, names and dates when new being as follows: No. 300 *Lyons*, September, 1876; No. 301 *Caen*, March, 1877; Nos 302 *Turin*, 303 *Milan*, 304 *Nice*, and 305 *Genoa*, December, 1877; and Nos 306 *Naples* and 307 *Venice*, March, 1878. When first brought out Nos 300 and 301 had the large standard tenders with outside frames, whilst Nos 302–307 had old tenders from Craven's engines. Later on, however, Nos 300 and 301 had old Craven tenders, their original ones being used on the last of Stroudley's new goods engines. Now, several of this class have the tenders of the old 'C' class whose engines have been scrapped. The weight of these locomotives is distributed as follows: 12 tons 1 cwt on the leading wheels, 15 tons on the driving and 7 tons 6 cwt on the trailing, total 34 tons 7 cwt. All the details and fittings are identical with those of the standard 'D' class passenger tanks, the whole of the motion, cylinders, wheels and side rods being interchangeable. They were formerly fitted with the speed indicator. In June, 1883, six more of this design were put to work on the passenger service owing to the good results obtained with the previous engines. These are numbered and named as follows: No. 308 *Como*, No. 309 *Splugen*, No. 310 *Laval*, No. 311 *Rhone*, No. 312 *Albion* and No. 313 *Paris*. The latter engine always works the 8.00 p.m. Grande Vitesse

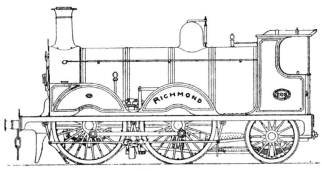

Fig. 131.

train from London to Newhaven. These engines differ from the former ones by having iron brake blocks on both sides of the coupled wheels only, and not on the trailing wheels, whilst at the bottom of the brake hanger is a small cylinder and piston, which is connected by a tube to the main reservoir, each wheel therefore having a separate cylinder for applying the air brake. The clack boxes are very low down on the boiler side and placed on the middle ring. The tenders are of the standard passenger engine design, with inside bearings. Nos 301 and 307 were scrapped in 1902.

Stroudley, having found that the front coupled engines were very satisfactory, resolved to build some express engines on the same principle, with larger wheels, for working the 8.45 a.m. Brighton to London express. They are known as the 'D 3' class. These engines, *Figs 131 and 132*, had 6 foot 6 inch leading and driving wheels, and have thoroughly stultified all the predictions made at the time of their being built, that their big leading wheels would throw them off the road, for they take curves at high speed with the greatest ease. The cylinders are 17½ inch diameter and the stroke of the pistons 26 inches. They are placed at an inclination of 1 in 11½, with their centres 2 feet 2 inches apart. The coupled wheels are 6 feet 6 inches in diameter and the trailing wheels 4 feet 6 inches. The wheelbase is 15 feet 7 inches, between leading and driving centres being 7 feet 7 inches and between driving and trailing 8 feet. The boiler is 4 feet 3 inches in diameter outside the largest ring and 10 feet 2 inches long, the centre line being 7 feet 5 9/16 inches above the rails. The length of the firebox casing is 5 feet 8¼ inches and the breadth 4 feet 1 inch, the depth being at the front end 5 feet 6 inches, and at the back end 5 feet. The heating surface of the tubes is 1,073.84 sq. ft and that of the firebox 109 sq. ft, giving a total of 1,182.84 sq. ft. The grate area is 17 sq. ft. These engines originally had 202 tubes 1¾ inch diameter, but now have 272 tubes 1½ inch diameter. The firebox roof and stays have also been altered as in previously described engines. The weight on the leading wheels is 13 tons 10 cwt, on the driving 14 tons 10 cwt and on the trailing 8 tons, giving a total of 36 tons. The clack boxes and delivery pipes are on the side of the boiler and the crown of the firebox casing slants towards the back, the fall being 2 inches. By comparing the two diagrams the difference between the brake gear of the first engine of this design and the remaining five will be noticed. A few slight differences also exist between the first engine and the others in the framing, the trailing footstep, the leading guard irons and the pumps. When new

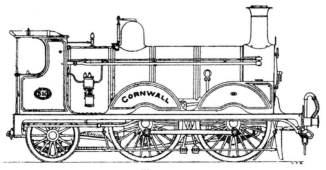

Fig. 132.

they all had the large goods tenders, but now have the standard inside-framed tenders, which will be described in connection with the 'G' class. The 'D 3' class worked the Brighton express until the 'Gladstones' came out, when three of them were transferred to Eastbourne to work the fast 8.30 a.m. Up and 5.05 p.m. Down, which service they performed until a few years back. Their numbers, names and dates were as follows: No. 208, *Richmond*, October, 1878; No. 209, *Devonshire*, July, 1879; No. 210, *Cornwall*, October, 1879; Nos 211, *Beaconsfield*, 212, *Hartington*, and 213, *Norfolk*, March, 1880. In 1885 No. 211 was renamed *Cavendish*. No. 210 *Cornwall*, has been fitted with the automatic vacuum brake for working through excursion trains, and in June, 1902, was re-named *Belgravia*. In August, 1897, Nos 208 and 209 were re-numbered 508 and 509; in September, 1897, Nos 210 and 211 were re-numbered 510 and 311; and in October, 1897, Nos 212 and 213 were re-numbered 512 and 513. These engines have again been re-numbered 608–613, as new engines were built to take the Nos 508 to 513. Nos 609 and 613 were scrapped in December, 1901, No. 611 in December, 1902, and No. 612 in April, 1903.

It was at the end of the year 1880 that Stroudley introduced his well-known 'G' class single driving wheel express engines. These have become famous for the excellent work that they have performed during the past twenty-two years, having for their size and weight proved second to none. Not a few times have they been called upon, in emergencies, to run the 5 p.m. Brighton express, and have arrived at Brighton to time. Until the new bogie expresses came out in 1895 the majority of the fast and heavy expresses to and from Portsmouth were worked by these engines, whilst they also ran the fast Eastbourne trains previous to the 'D 3' class being stationed at that town. They are illustrated by *Fig. 133*, The cylinders, which are horizontal, are 17 inches by 24 inches, their centres being 2 feet 2 inches apart, the travel of the valves is 3¾ inches, the outside lap ⅞ inch the inside lap ⅛ inch, and the lead 1/16 inch; the diameter of the blast pipe orifice is 4¼ inches, and its height above the centre line of the boiler 1 foot 2½ inches. The diameter of the leading and trailing wheels is 4 feet 6 inches and of the driving 6 feet 6 inches. The overhang at the leading end is 5 feet 8¼ inches and at the trailing end 3 feet 3¼ inches; the wheelbase is 15 feet 11 inches, the leading and driving centres being 8 feet apart, and the driving and trailing 7 feet 11 inches. The frames, which are of steel, are ⅞ inch thick. The boiler is of the standard design, in three rings, the external diameters being 4 feet 1 inch, 4 feet 2 inches, and 4 feet 3 inches, and the

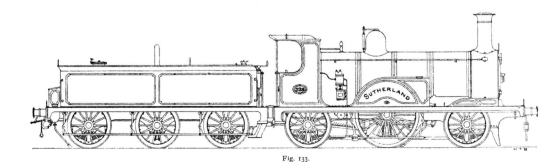

Fig. 133.

length of the barrel 10 feet 2 inches. There are 262 tubes 1½ inch diameter and 10 feet 8¼ inches long. The firebox casing is 5 feet 8¼ inches long, 4 feet 1 inch wide, and has a depth of 5 feet 2 inches below the centre line of the boiler at the front, and of 4 feet 8 inches at the back. Originally the inside box was stayed to the outer shell with rigid stays, but these have now been replaced by girder roof stay bars. The crown of the inside box is lower at the back end than at the front, but the crown of the outer shell is parallel to the centre line. The chimney is 13 feet 2 inches above the rail level, and the centre line of the boiler 7 feet 1 7/16 inches. The working pressure was formerly 140 lbs per sq. in., but is now 150 lbs. The heating surface of the tubes is 1,084.51 sq. ft, and that of the firebox 99.80 sq. ft, giving a total of 1,184.31 sq. ft. The grate area is 17 sq. ft. The weight of these engines in working order is as follows: 12 tons on the leading wheels, 13 tons 10 cwt on the driving, and 7 tons 18 cwt on the trailing, giving a total of 33 tons 8 cwt.

The tender, with 2 tons of coal and 10 tons of water, weighs 27 tons 7 cwt. The wheelbase of engine and tender is 38 feet 1½ inches, and the total length over the buffers is 50 feet 11 inches. The tender is of the well-known standard type, having inside frames and wheels 4 foot 6 inch diameter. The latter are interchangeable with the leading wheels of the 'G' class, and the trailing of the 'G', 'D', 'D 2', 'D 3' and 'B' classes. The capacity of the tank is 2,250 gallons of water, and the coal space will accommodate about 4 tons. The wheelbase of the tender is 14 feet, equally divided. The exhaust steam from the cylinder is admitted into the tank so as to heat the feed water. The clack boxes and pumps are similar to those of the latest 'D 2' class. These engines originally had Stroudley's speed indicator.

All this class are fitted with the Westinghouse automatic air brake, and No. 341 *Parkhurst* has in addition the automatic vacuum brake apparatus. *Stephenson* went in 1881 to the Stephenson Centenary, at Newcastle, and took part in the procession of engines at Wylam.

They were all built at the Brighton Works, and the names, numbers and dates are:

No.	Name	Date	No.	Name	Date
327	Imberhorne	Dec., 1880	339	London	Dec., 1881
328	Sutherland	"	340	Medina	"
329	Stephenson	June, 1881	341	Parkhurst	"
330	Newhaven	"	342	St Lawrence	"
331	Fairlight	"	343	Wilmington	"
332	Shanklin	"	344	Hurstmonceux	"
333	Ventnor	"	345	Plumpton	Mar., 1882
334	Petworth	"	346	Alfriston	"
335	Connaught	Sept., 1881	347	Dallington	"
336	Edinburgh	"	348	Lullington	April, 1882
337	Yarmouth	"	349	Albany	"
338	Bembridge	Oct., 1881	350	Southbourne	"

The next type Stroudley introduced was a very powerful goods engine, the new 'C' class or 'Jumbo's', as they are more generally known. There are twelve altogether, the first appearing in June, 1882, and the last in June, 1887, *(Fig. 134)*. The cylinders are 18¼ inch diameter, and the stroke of the pistons 26 inches; the centres are 2 feet 1 inch

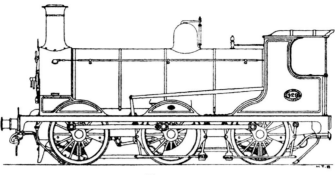

Fig. 134.

apart. They are placed at an inclination of 1 in 11½, the valves, which are underneath, being inclined at an angle of 1 in 15. The steam ports of these cylinders, which are both made in one casting, are 1⅜ inches wide and 15 inches long, the exhaust ports being 2 inches by 15 inches. The blast pipe orifice is 5 inch diameter, and the height above the centre line of the boiler 12⅞ inches. The overhang at the leading end is 5 feet 1½ inches, and at the trailing end 5 feet 1½ inches, and at the trailing end 5 feet 10¼ inches; the wheel-base is 15 feet 3 inches, being 7 feet 9 inches between the leading and driving wheels, and 7 feet 6 inches between the driving and trailing. The diameter of the wheels is 5 feet. The frames are of steel and 1 inch thick. The boiler is in three rings, the diameters being 4 feet 5¾ inches, 4 feet 6⅞ inches, and 4 feet 8 inches, and its length is 10 feet 2 inches. The length of the firebox casing is 6 feet 8¼ inches, the width 4 feet 1 inch, and the depth at the front below the centre line of the boiler 5 feet and at the back 3 feet 4 inches, the trailing axle being underneath. The boiler contains 317 tubes 1½ inches in diameter by 10 feet 8¼ inches long, and its centre line is 7 feet 1 inch above the rails. The working pressure is 150 lbs per sq. in. The heating surface of the tubes is 1,312 sq. ft, and that of the firebox 101 sq. ft, giving a total of 1,413 sq. ft. The grate area is 20.95 sq. ft. Length of the engine and tender over the buffers is 48 feet 7 inches. The weight on the leading wheels is 13 tons 14 cwt, on the drivers 14 tons, and on the trailers 12 tons 13 cwt, total 40 tons 7 cwt, a comparatively light weight for engines of such ample dimensions, lightness, as will be noticed, being one of Stroudley's chief features. The tenders for these engines were of the large standard type with outside bearings, particulars of which have already been given with the description of *Goodwood* and *Belgravia*. The 'C' class are numbered from 421–432; they were all built at Brighton at the following dates:

No.	Date	No.	Date
421	June, 1882	427	May, 1884
422	”	428	”
423	Sept., 1882	429	”
424	”	430	June, 1887
425	Oct., 1882	431	”
426	Nov., 1882	432	”

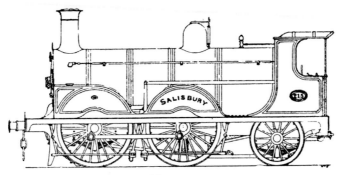

Fig 135.

The last six have an air reversing cylinder fixed to the reversing screw and wheels, and the reversing rod is outside the firebox lagging. In the first six it is between the lagging and the firebox, coming out at the front end. These engines have the Westinghouse air brake with duplex pumps employed for feeding the boilers, similar to those of the 'E' class. In March, 1897, No. 431 was fitted with Smith's patent piston valves. These engines have the tenders of the 'D 3' class, *Grosvenor* and *Abergavenny* and two 'D 2', or ten tenders between twelve engines. They were originally fitted with Stroudley's speed indicator.

The passenger train service on the LBSCR had for a long time been steadily increasing, and the trains becoming more numerous, heavier, and having to be run at higher speeds, locomotives had to be provided of increased dimensions to meet the demand for more power. Accordingly the famous *Gladstone*, the first engine of the 'B' class, was designed in 1881. It was not, however, put in hand then, and was not finished before December, 1882. An engine of this class is illustrated by *Fig. 135*. The cylinders are cast together in one piece, and are 18¼ inch diameter by 26 inch stroke – Stroudley was the first to adopt cylinders cast in one for locomotives – the distance from centre to centre being only 2 feet 1 inch, thus permitting the webs of the crank to be made 5 inches thick, and an inside axle bearing 8 inches long, whilst still leaving plenty of room for the eccentric sheaves. The cylinders are placed at an inclination of 1 in 11½ inches, and the length of the ports is 15 inches, the width of the steam ports being 1⅜ inches, and of the exhaust ports 2 inches. The slide valves are underneath the cylinders at an angle of 1 in 15. The centre line of the boiler is 7 feet 5 inches from the rails, the length of the barrel 10 feet 2 inches, diameter outside 4 feet 6 inches, thickness of plates ½ inch, length of firebox shell 6 feet 8¼ inches, breadth of firebox outside at bottom 4 feet 1 inch, depth below boiler centre line at the front 5 feet 6 inches, and at the back 4 feet 3 inches. The water spaces at the bottom are 3 inches wide, tapering upward to about double at the top, the side plates being pocketed forward so as to enlarge the tube plate. There are 331 tubes, which are of steel, the length over all being 10 feet 8¼ inches, diameter outside 1½ inches; diameter of blast pipe orifice 4¾ inches. The heating surface of the tubes is 1,372.92 sq. ft, of the firebox 112.48 sq. ft, total 1,485.40 sq. ft. The grate area is 20.65 sq. ft. The diameter of the leading and driving wheels is 6 feet 6 inches, and that of the trailing wheels 4 feet 6 inches, the distance between the centres of the leading and driving wheels being 7 feet 7 inches, and between the driving and trailing wheels 8 feet; from driving wheel centre to front of firebox is 1 foot 10⅜ inches, the leading overhang is 5 feet 10 inches, and the trailing overhang 4 feet 4 inches. The weight of the engine in working order is: on leading wheels 13 tons 16 cwt, on driving wheels 14 tons 10 cwt, and on trailing wheels 10 tons 8 cwt, total 38 tons 14 cwt. The numbers, names and dates of these engines are:

No.	Name	Date	No.	Name	Date
172	Littlehampton	Apr., 1891	190	Arthur Otway	Dec., 1888
173	Cottesloe	"	191	Gordon-Lennox	Nov., 1888
174	Fratton	Dec., 1890	192	Jacomb-Hood	Oct., 1888
175	Hayling	"	193	Fremantle	"
176	Pevensey	Nov., 1890	194	Bickersteth	June, 1888
177	Southsea	"	195	Cardew	"
178	Leatherhead	June, 1890	196	Ralph L. Lopes	May, 1888
179	Sandown	May, 1890	197	Jonas Levy	"
180	Arundel	Mar., 1890	198	Sheffield	Dec , 1887
181	Croydon	Feb., 1890	199	Samuel Laing	"
182	Hastings	Dec., 1889	200	Beresford	"
183	Eastbourne	Nov., 1889	214	Gladstone	Dec , 1882
184	Carew D. Gilbert	Sept,1889	215	Salisbury	Dec., 1883
185	George A. Wallis	"	216	Granville	"
186	De la Warr	June, 1889	217	Northcote	"
187	Philip Rose	"	218	Beaconsfield	Oct., 1885
188	Allen Sarle	Apr., 1889	219	Cleveland	"
189	Edward Blount	Mar., 1889	220	Hampden	Dec., 1887

No. 172 *Littlehampton* differs slightly from the rest of the class in having duplex injectors placed on the firebox front instead of pumps, and also by not having pockets in the side plates of the inside firebox, whilst the tubes number 265, and are 1⅝ inch diameter. No. 175 *Hayling* was the engine that ran the 8.45 a.m. from Brighton on 1 May, 1891, when the Portland Road bridge at Norwood Junction collapsed, throwing the whole of the train off the line. No. 195 *Cardew* usually runs the 9 p.m. boat train from London Bridge to Newhaven. No. 189 *Edward Blount* was sent to the Paris Exposition of 1889, where it obtained a gold medal, and after the close of the great Exhibition it remained in France for a few weeks to run trials on the Paris, Lyons & Mediterranean Railway in competition with the SER's engine No. 240. It was in connection with these experiments that Stroudley contracted the chill that proved fatal in December, 1889. The trials took place on the main line of the PLM, between Paris and Laroche. The high speed trials were made for the first time on 17 January, 1890, and again on the 18th. They took place between Montereau and Sens. The trial of drawing a heavy train was repeated twice – on 19 December, 1889, and 15 January, 1890. The run was from Paris to Laroche and back. In the later engines of this class a few alterations have been made as experience suggested an improvement, notably in the steam chest, the length of the steam ports being increased from 15 inches to 16 inches, and the width from 1⅜ inches to 1½ inches, this being done to enable the engine to get rid of the steam more freely with heavy loads. Another alteration is in the reversing gear, by improving the air cylinder for reversing the engine. Adams'

annular blast pipe has also been substituted for the ordinary one, the boiler pressure has been increased from 140 lbs to 150 lbs per sq. in., and the number of tubes from 331 to 333. The front edge of the cab of No. 214 follows the curve of the trailing wheel at the bottom similarly to that of the *Grosvenor*, but in all other engines of this class it terminates in one large radius as shown in the diagram. Some of these engines had a special kind of balanced slide valve designed by Stroudley, and some also had a pneumatic arrangement for locking the air reversing gear; this consisted of a shallow square chamber containing a piston which was forced against the guides of the reversing nut on the admission of air from the Westinghouse brake through a flexible steel pipe. Both these arrangements, however, proved unsatisfactory, and were soon removed. All the engines are fitted with the Westinghouse automatic air brake, and the following have in addition the apparatus for working the automatic vacuum brake: Nos 175, 176, 181, 187, 192, 194 and 197. Nos 178 and 190 are fitted with Macallan's patent variable blast pipe, and Nos 198, 200 and 215 with apparatus for heating the carriages with steam. A few years ago No. 186 went through with a special from Brighton to Birmingham *via* Addison Road, and thence over the LNWR. Nos 186 and 198 have now been fitted to burn liquid fuel on Holden's system.

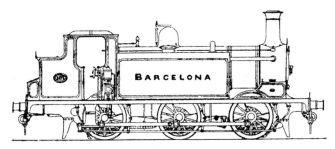

Fig. 136.

In October, 1884, Stroudley built at Brighton an extra powerful goods tank engine, designed specially for working goods trains over the Eastbourne and Tunbridge Wells line, which abounds in severe gradients and sharp curves all the way from Hailsham to Eridge. This locomotive is numbered 157, and named *Barcelona*, (*Fig. 136*). The cylinders are Stroudley's standard 18¼ inch by 26 inch design, with the valves underneath. The wheels are 4 feet 6 inches in diameter as with the other goods tank engines, with which also the wheel-base is identical. The boiler is of the same dimensions as those of the 'G' class express engines, with the exception of the length of the firebox, which is made the same as those of the 'E' class tanks, owing to the wheelbase being similar. The heating surface is 1,171.51 sq. ft, of which the tubes give 1,084.51 sq. ft and the firebox 87 sq. ft. The side tanks are much higher and carry 1,052 gallons of water, and the coal bunker is much larger also. Instead of the motion being reversed by a lever as on the other goods tanks, it is operated by a wheel and screw, as on the passenger engines. The clack boxes are out of sight, being placed under the boiler behind the tanks. The boiler is higher, its centre line being 7 feet 2 15/16 inches from the rail. The *Barcelona* is fitted with the Westinghouse brake with duplex pump for feeding the boiler, in addition to the pumps worked from the crosshead. Its weight in working order is 43 tons 18 cwt, the leading wheels carrying 14 tons 1 cwt, the driving 15 tons 9 cwt, and the trailing 14 tons 8 cwt. Amongst other minor external differences are the larger leading sandboxes, the absence of a handrail on the sides of the boiler, and the small handrail placed on the front of the side tank. This engine also was fitted with the air reversing gear and pneumatic locking gear. It was originally painted the standard passenger engine colour. A few years ago No. 157 was removed to Brighton and an eight-wheeled goods tank put on the Eastbourne and Tunbridge Wells service, but owing to the sharp curves this engine was not found satisfactory, and No. 157 was sent back again.

On Friday, 20 December, 1889, news reached England that William Stroudley had died at the Grand Hotel, St Lazare, Paris, where he had been staying for the purpose of conducting the engine trials just mentioned. During these he contracted a bad cold, resulting in a severe attack of asthma, from which he never recovered.

William Stroudley

Stroudley was born on 6 March, 1833, at Sandford, Oxfordshire, and in 1847 was apprenticed to John Inshaw, of Birmingham. In 1848 he went to work a stationary engine at the Vulcan Foundry, Birmingham, and later performed similar duties for a Mr Dean, also in Birmingham. In 1849 he was again employed under Inshaw, working on the engines and boilers of some twin-screw boats for the Grand Canal Company, of Dublin. In 1851 he was working at the Islington Foundry, Birmingham, on stationary engine work; and in May, 1853, was engaged at the GWR Works, Swindon, under Daniel Gooch. In 1854 be went to the Peterborough Shops of the Great Northern Railway as running shed fitter; and in 1857 took charge of a small line of railway for Lord Willoughby d'Eresby. In 1858 he was employed at the Helpstone Paper Mills, near Peterborough, having charge of the engines and machinery, and in 1859 was again at Peterborough on his old job as shed fitter. In October, 1861, Stroudley was made works manager of the Cowlairs shops of the Edinburgh & Glasgow Railway, and in 1865 was appointed locomotive and carriage superintendent of the Highland Railway at Inverness. In January, 1871, he was appointed locomotive and carriage superintendent of the LBSCR. On 24 May, 1870, he was made Assoc. M.I.C.E., and a full member on 6 February, 1877. At the time of his death Stroudley had under construction or on order eighteen engines. There were being built eleven engines of the 'B' class, Nos 172–182, and a large six wheels coupled radial tank engine, while material was also under order for six standard goods tank engines. The 'B' class was finished entirely to Stroudley's designs, except for the minor alterations in the last engine, No. 172, which have already been mentioned.

Chapter 5

Robert J. Billinton, 1890–1903

In February, 1890, Robert J. Billinton was appointed locomotive, carriage and wagon superintendent. He had previously held the responsible position of chief draughtsman of the locomotive department of the Midland Railway at Derby, while he had before that held a similar position at the Brighton Works.

It was not until December, 1891, exactly two years after the death of Stroudley, that the large tank engine already referred to was finally completed. The boiler was originally put in hand during Stroudley's lifetime, and the engine was built at odd times during the two years, it practically being looked upon as a stock job. Billinton introduced into the design a few minor alterations of his own, substituting for the tool-box, which would have been at the back of the coal bunker, one on the footplate inside the cab, the chief advantage being that the bunker could then hold more coal and a larger tank could be put beneath it. A cast-iron chimney was substituted for the wrought-iron one with polished copper cap, so familiar on all Brighton engines. The clack-boxes were dispensed with from the underside of the boiler, and Gresham & Craven's combination injector and clack-box placed on the firebox front instead. Cast-iron splashers were used for the leading wheels and cast-iron sandboxes, both in front and behind the driving wheels, these being placed under the footplate and fixed to the

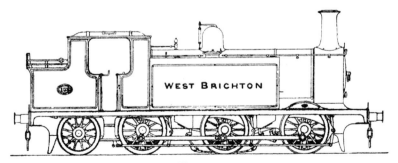

Fig. 137.

framings. Owing to the employment of injectors, cold water is used for feeding the boiler. This, therefore, necessitates the cutting-off from the tank of the exhaust steam from the cylinders. In April, 1899, this engine had the cast-iron chimney replaced by a standard 'E' class funnel with polished copper cap. A few minor alterations were also made in the radial axle-box gear. A sliding ventilator was added to the cab roof, and the cab windows were hinged so as to open, the brass ring round the window on the outside of the cab being dispensed with. It should be mentioned that when the cast chimney was employed a cast-iron liner was also used, tapered the opposite way to the outside, being larger at the top. Another special feature of this engine was the introduction of cast-steel wheels with balance weights of a half-moon shape. This was decided upon by Stroudley. It might here be noted that a good many engines of the 'G' and 'B' classes have had their original wrought-iron wheels replaced by others of cast-steel with the same shaped counterweights.

Returning to the goods tank, this engine, *Fig. 137*, is officially described as 'E class special'. It is numbered 158 and named *West Brighton*, is stationed at Fratton and works between Portsmouth and Brighton. The cylinders are the standard 18¼ inch by 26 inches type, with the slide valves underneath. The coupled wheels are 4 feet 6 inches in diameter and the radial trailing pair 4 feet. From the front of the leading buffer-plate to the centre of the leading wheels is 6 feet 0¼ inch. The leading and driving centres are 7 feet 6 inches apart, the driving and intermediate centres 7 feet 9 inches and the intermediate and trailing centres 6 feet, the total wheel-base being 21 feet 3 inches; the overhang at the trailing end is 4 feet 7¼ inches. The radial axle-box is of Stroudley's design, but improved upon by the present locomotive superintendent; its radius is 13 feet 0⅜ inches. The boiler is of the standard design, the external diameters being 4 feet 1 inch, 4 feet 2 inches and 4 feet 3 inches and the length of the barrel 10 feet 2 inches. The crown of the inside firebox slants towards the trailing end. The length of the outside casing is 5 feet 8¼ inches, the depth below the centre line at the front end being 5 feet 2 inches and at the trailing end 4 feet 2 inches. The boiler contains 262 tubes 1½ inch diameter and 10 feet 8¾ inches long; its centre line is 7 feet 4 5/16 inches above the rail level. The heating surface of the tubes is 1,082.22 sq. ft, and of the firebox 91.70 sq. ft, giving a total of 1,173.92 sq. ft. The area of the firegrate is 17 sq. ft. The link motion is reversed by means of a wheel and screw. This engine is not fitted with the Westinghouse air brake, but with a powerful steam brake which acts

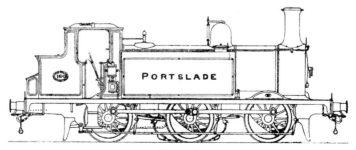

Fig. 138.

on the coupled wheels. The capacity of the side tanks is 1,055.52 gallons, that of the bunker tank 343.75 gallons, and of the three connecting pipes 16.50 gallons, giving a total capacity of 1,415.77 gallons. The coal bunker will carry tons of coal; railings, which were not in the original design, being carried round it. The weight of the engine is 51 tons 14 cwt, distributed as follows: 12 tons 17 cwt on the leading wheels, 15 tons 8 cwt on the driving wheels, 12 tons 8 cwt on the intermediate wheels and 11 tons 1 cwt on the trailing wheels.

The next class of engine is that of the smaller goods tanks, which underwent still further alterations from the intended design. They are practically the same as the standard 'E' class tanks, and as far as dimensions go, they are identical. The chief alterations are in external appearance and in the boilers which are made from the drawings of Billinton. Illustrated by *Fig. 138*, and are numbered and named as:

No.	Name	Date	No.	Name	Date
159	Edenbridge	June, 1891	162	Southwater	Nov., 1891
160	Portslade	"	163	Southwick	Dec., 1891
161	Aldrington	Nov., 1891	164	Spithead	"

As in the case of *West Brighton*, these engines had cast-steel wheels, cast-iron chimneys, splashers and sandboxes, but are fitted with the Westinghouse air brake. They also have the sliding ventilator in the roof, and cab windows made to open; combination injectors and exhaust up the chimney. The wheels, cylinders, wheel-base and motion are identical with the standard 'E' class. The boiler differs in details, the dome, which is of wrought iron, being on the middle plate, and a manhole is provided on the firebox. The roof stay bars are of steel plate girders, each stay being made up of two plates rivetted together with thick washers between them. The weight of these engines is 41 tons 8 cwt, the leading wheels carrying 13 tons 9 cwt, the driving 15 tons and the trailing 12 tons 19 cwt. They are fitted, as is No. 158, also with Gresham & Craven's steam sanding gear. The cast-iron chimneys have now been replaced by wrought-iron ones with copper tops.

In 1891 an entirely new type of passenger tank engine was designed for the LBSCR, having four wheels coupled and a trailing bogie. The first of these was built in May,

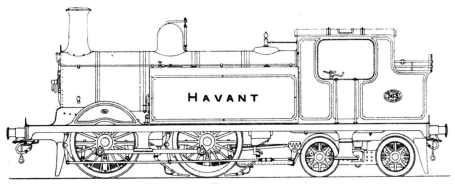

Fig. 139.

1892, and is known as 'D bogie' class; *(Fig. 139)*. These engines are used for all classes of passenger work, from fast main-line trains to locals, and are numbered 363–398. The cylinders, which are inclined at an angle of 1 in 9½, are 18 inches in diameter, with a stroke of 26 inches, and are placed with their centre line 10 feet 3½ inches from the centre of the crank axle. The slide valves are placed between the cylinders, but in order to gain room and avoid a cramped steam chest, they are at an angle of 1 in 7¾. The valve motion is similar to Stroudley's, but has two balance weights on arms solid with the weigh-bar shaft. It is reversed by a wheel and screw. The leading and driving wheels are 5 feet 6 inches in diameter and the bogie wheels 3 feet. From the front of the leading buffer beam to the centre of the leading axle is 6 feet 1½ inches, between leading and driving centres 8 feet, from the centre of the driving axle to the centre of the bogie 11 feet 10 inches, between centres of the bogie wheels 6 feet, and from the centre of the trailing pair of bogie wheels to the trailing buffer beam 2 feet 7½ inches. The total length over the buffers is 34 feet 7 inches. The boiler centre line stands 7 feet 4 5/16 inches, from the rail level, the barrel being in three plates, having external diameters of 4 feet 1 inch, 4 feet 2 inches and 4 feet 3 inches respectively; its length from the back of the front tube plate to the front of the throat plate is 10 feet 4¼ inches. The front tube plates of all Billinton's engines are circular and are let into the barrel of the boiler, the latter resting upon a cast-steel bracket fixed to a flange at the back of the cylinders. The smokebox is double and the chimney has a liner tapering the opposite way to the outer chimney; that is, smaller at the bottom than the top. The length of the firebox casing is 5 feet 8¼ inches, the breadth 4 feet 1 inch and the depth below the centre line of the boiler 5 feet 2 inches at the front and 4 feet 8 inches at the back. The crown of the firebox is parallel to the outer shell, from which the eight girder stays securing it are slung. The barrel contains 242 tubes of 1⅝ inch diameter, giving a heating surface of 1,106.44 sq. ft, while that of the firebox is 99 sq. ft, making a total of 1,203.44 sq. ft. The grate area is 17.083 sq. ft. The side tanks carry 884.86 gallons, the tank under the coal bunker 262.25 gallons and three connecting pipes 14.3 gallons, giving a total of 1,161.41 gallons, whilst the bunker has a capacity for 2 tons of coal. The weight in working order is 48 tons 9 cwt, of which 14 tons 17 cwt is on the leading wheels, 16 tons 13 cwt on the driving wheels and 16 tons 19 cwt on the bogie. The working pressure of the boiler is 160 lbs per sq. in. These engines are fitted with Gresham & Craven's steam sanding gear and duplex injectors. All of them, of course, have the Westinghouse air brake, whilst Nos 367–374 and 397 are also fitted for working the

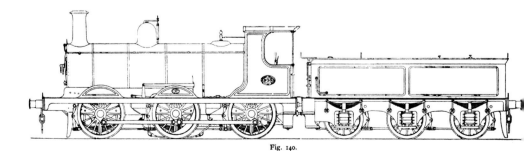

Fig. 140.

automatic vacuum brake. They are the only engines of Billinton's design that have outside brake-rods.

All the engines of this class have been built at Brighton, their numbers, names and dates being as follows:

No.	Name	Date	No.	Name	Date
363	Goldsmid	May, 1892	381	Fittleworth	Oct., 1893
364	Truscott	”	382	Farlington	”
365	Victoria	June, 1892	383	Three Bridges	Dec., 1893
366	Crystal Palace	”	384	Cooksbridge	”
367	Norwood	”	385	Portsmouth	Jan., 1894
368	Newport	”	386	Chichester	”
369	Burgess Hill	Oct, 1892	387	Steyning	May, 1894
370	Haywards Heath	”	388	Emsworth	”
371	Angmering	Dec., 1892	389	Shoreham	”
372	Amberley	”	390	St. Leonard's	”
373	Billingshurst	Jan., 1893	391	Drayton	June, 1894
374	Pulborough	”	392	Polegate	”
375	Glynde	Apr, 1893	393	Woodside	Mar., 1896
376	Folkington	”	394	Cowfold	”
377	Hurstpierpont	June, 1893	395	Gatwick	Apr., 1896
378	Horsted Keynes	”	396	Clayton	”
379	Sandersted	”	397	Bexhill	Nov., 1896
380	Thornton Heath	”	398	Haslemere	”

In September, 1895, the name of No. 363 was changed to *Havant*.

The next type of engine designed at Brighton was the standard six-coupled goods, known as class 'C'. All the engines of this class have been built by the Vulcan Foundry Co., of Warrington, in 1893, 1894, and 1900-1-2; they are numbered 433–452 and 521–550, *(Fig. 140)*. The valve gear and cylinders are the same as in the 'D bogie' class, and are placed at the same inclination. The motion is reversed by a lever. The wheels are 5 feet in diameter, the wheelbase being 16 feet equally divided. The overhang at the leading end is 6 feet 1½ inches, and at the trailing end 4 feet 7 inches. The boiler barrel is exactly the same as in the bogie tank engines, but its centre is 7 feet 4⅜ inches from the rails. The length of the outside firebox is 6 feet 2¼ inches, the width 4 feet 1 inch, the depth 5 feet 2 inches at the front and 3 feet 9¾ inches at the back. The heating surface of the firebox is 105.25 sq. ft, and that of the tubes 1,106.44 sq. ft, giving a total of 1,211.69 sq. ft. The area of the fire grate is 19.322 sq. ft. The weight, empty, is 35 tons, and in working order 38 tons 12 cwt, distributed as follows: on leading wheels, 13 tons 7 cwt, on driving wheels 12 tons 15 cwt, and on trailing wheels 12 tons 11 cwt. The tender is carried on six wheels 4 feet in diameter, on a base of 13 feet, equally divided. The leading overhang is 4 feet 4 inches and the trailing 3 feet 10½ inches. The capacity of the tank is 2,420 gallons, and the coal carried is about 4 tons. The total length of engine and tender over buffers is 51 feet 5 inches. The weight of the tender in full working order is 32 tons 13 cwt, making a total for engine and tender 71 tons 5 cwt. All the engines are fitted with the Westinghouse air brake, and during the summer are used a good deal on heavy passenger trains, running quite steadily at 55 miles per hour. Nos 433–444 (maker's Nos 1375–1386) were built in June, 1893, and

Nos 445–452 (maker's Nos 1412–1419) in December, 1894; Nos 521–523 (maker's Nos 1699–1701) were built in August, 1900; Nos 524–527 (makers' Nos 1702–1705) were built in September; Nos 528–534 (maker's Nos 1706–1712) were built in October; Nos 535–540 (maker's Nos 1713–1718) were built in November, 1900; No. 541 (maker's No. 1813) was built in November, 1901; Nos 542–546 (maker's Nos 1814–1818) were built in December, 1901; Nos 547–550 (maker's Nos 1819–1822) were built in January, 1902. The maker's name-plates on engines Nos 433 and 434 are on the side of the driving splashers as well as at the back of the tender, but the remaining engines have them at the back of the tender only.

In 1894 drawings were prepared for some six-wheels coupled radial tank engines of the 'West Brighton' type, but entirely from Billinton's design, and the first of this class, –*Fig. 141)*, appeared in November of that year. They had the standard 18 inch by 26 inch cylinders placed at an inclination of 1 in 9½, with the slide valves inclined at the rate of 1 in 7¾. The coupled wheels are 4 feet 6 inches in diameter, and the radial trailing wheels 4 feet. The overhang at the leading end is 6 feet 1½ inches, and the total wheel-base 21 feet 6 inches, being from leading to driving centres 7 feet 9 inches, from driving to intermediate 7 feet 6 inches, and from intermediate to trailing 6 feet 3 inches. The overhang at the trailing end is 4 feet 7½ inches. The boiler is of the standard type, 4 feet 3 inches in diameter outside the largest ring, and 10 feet 7¼ inches long. Like all the other engines of Billinton's design, they had cast steel brackets attached to the back of the cylinders, upon which the barrel of the boiler takes a bearing. The centre line of the boiler is 7 feet 5⅜ inches from the rails. The length of the firebox casing is 5 feet 8¼ inches, the breadth 4 feet 1 inch, the depth below the centre line of the boiler at the front 5 feet 2 inches, and at the back 4 feet 1¾ inches. The boiler contains 242 brass tubes of 1⅝ inch diameter, with a heating surface of 1,106.44 sq. ft, whilst that of the firebox is 93 sq. ft, giving a total of 1,199.44 sq. ft. The capacity of the side tanks is 1,020 gallons, these being much higher than those of the bogie tank engines. The tank under the bunker carries 338 gallons, and the three connecting pipes 19 gallons, giving a total water capacity of 1,377 gallons. The coal bunker carries 2½ tons. The area of the fire grate is 17.43 sq. ft, and the area through the tubes 2.65 sq. ft. The total length over buffers is 35 feet 3 inches, and the weight in working order 51 tons, of which 13 tons is on the leading wheels, 14 tons on the drivers, 11 tons on the intermediate

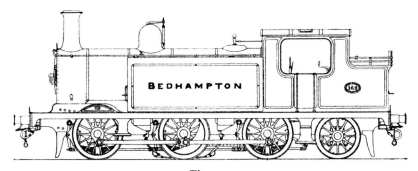

Fig. 141.

wheels, and 13 tons on the trailers. All the sixteen engines of this class are fitted with the Westinghouse automatic brake, and are very extensively used on passenger trains. Their numbers, names and dates are as follows:

No.	Name	Date	No.	Name	Date
165	Blatchington	Nov., 1894	455	Brockhurst	May, 1895
166	Cliftonville		456	Aldingbourne	"
167	Saddlescombe		457	Watersfield	Oct., 1895
168	Southborough	"	458	Chalvington	"
169	Bedhampton	Dec., 1894	459	Warlingham	"
170	Bishopstone	"	460	Warminghurst	Nov., 1895
453	Broadbridge	May, 1895	461	Staplefield	"
454	Storrington	"	462	Washington	"

These engines being so frequently used on passenger work, it was decided to build all engines of this type for the future with 6-inch larger wheels. Up to the present sixty-nine of them have been built, all at Brighton, their numbers, names and dates being:

No.	Name	Date	No.	Name	Date
463	Wivelsfield	Dec. 1897	498	Strettington	April, 1900
464	Woodmancote	"	499	Woodendean	May, 1900
465	Hurst Gren	Mar., 1898	500	Puttenham	"
466	Honor oak	"	501	Stoat's Nest	June, 1900
467	Berwick	April, 1898	502	Ridgewood	"
468	Midhurst	"	503	Buckland	Aug., 1900
469	Beachy Head	May, 1898	504	Chilworth	"
470	East Hoathly	"	505	Annington	Sept., 1900
471	Forest Hill	June, 1898	506	Catherington	Oct., 1900
472	Fay Gate	"	507	Horley	Nov., 1900
473	Birch Grove	July, 1898	508	Bognor	"
474	Bletchingly	"	509	Southover	Dec., 1900
475	Partridge Green	Sept., 1898	510	Twineham	"
476	Beeding	"	511	Lingfield	Jan., 1901
477	Poynings	Oct., 1898	512	Kingswood	Feb., 1901
478	Newick	"	513	Densworth	"
479	Bevendean	Nov., 1898	514	Barcombe	Mar., 1901
480	Fletching	"	515	Swanmore	"
481	Itchingfield	Dec., 1898	516	Rustington	April, 1901
482	Newtimber	"	517	Limpsfield	May, 1901
483	Hellingly	Feb., 1899	518	Porchester	"
484	Hackbridge	Mar., 1899	519	Portfield	June, 1901
485	Ashington	"	520	Westbourne	"
486	Godalming	April, 1899	556	Tadworth	Aug., 1901
487	Fishergate	May, 1899	557	Northlands	"
488	Oakwood	"	558	Chiltington	Sept., 1901
489	Boxgrove	June, 1899	559	Framfield	Oct., 1901
490	Bohemia	"	560	Pembury	"
491	Hangleton	Sept., 1899	561	Walberton	Nov., 1901
492	Jevington	"	562	Laughton	Dec., 1901
493	Telscombe	Oct., 1899	563	Wineham	"
494	Woodgate	"	564	Nettlestone	"
495	Chessington	Nov., 1899	565	Littleton	Jan., 1902
496	Chiddingfold	"	566	Durrington	Feb., 1902
497	Dennington	April, 1900			

Engine No. 517 has no safety valves on the dome, but 'Pop' safety valves on the firebox instead. Engines No. 518 and those following have the safety valves on the firebox and are of Billinton's design. Of these, a description will be found further on. It being intended to use these engines exclusively on passenger work, they have, commencing with No. 487 *Fishergate*, been painted the standard passenger engine colour.

The first of this new class is illustrated by *Fig. 142*. They are almost identical with their predecessors, but the boiler centre has been raised 3 inches, to 7 feet 8⅜ inches above the rails, making it at the time the highest pitched boiler of any tank engine in England. The side tanks only carry 1,000 gallons owing to the splashers being larger, but the tank under the bunker carries more, its capacity being 390 gallons, the connecting pipes carry the same, so that the total capacity is 1,409 gallons; the coal bunker is also longer and will carry 3 tons of coal.

In 1893 a type of locomotive entirely new to the LBSCR was designed at Brighton, *viz.*, the now well-known four-coupled bogie express engines. The first of the type did not appear, however, until June, 1895, being greatly improved upon from the original designs. It is illustrated by *Fig. 143*. The following is a complete list of the numbers, names, and dates of this class:

No.	Name	Date	No.	Name	Date
314	*Charles C. Macrae*	June, 1895	202	*Trevithick*	Feb., 1897
315	*Duncannon*	”	203	*Henry Fletcher*	Mar., 1897
316	*Goldsmid*	”	204	*Telford*	”
317	*Gerald Loder*	June, 1896	205	*Hackworth*	April, 1897
318	*Rothschild*	”	206	*Smeaton*	”
319	*John Fowler*	Sept., 1896	207	*Brunel*	June, 1897
320	*Rastrick*	”	171	*Nevill*	”
321	*John Rennie*	”	208	*Abercorn*	Aug., 1897
322	*G. P. Bidder*	Oct., 189	209	*Wolfe Barry*	Sept., 1897
323	*William Cubitt*	Dec., 1896	210	*Fairbairn*	Oct., 1897
324	*John Hawkshaw*	”	211	*Whitworth*	Nov., 1897
201	*Rosebery*	Jan., 1897	212	*Armstrong*	Jan., 1898

These locomotives were designed for replacing the 'G' class singles on the London and Portsmouth service, and afterwards on various other portions of the system.

Only three of these engines were built at first, and it was not until twelve months after

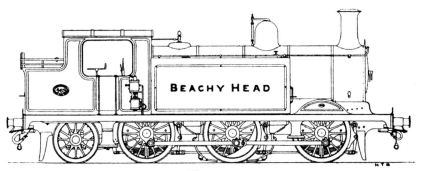

Fig. 142.

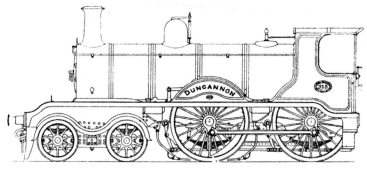

Fig. 143.

that the type was generally adopted. The cylinders are 18 inches in diameter by 26 inch stroke, and placed at an inclination of 1 in 20. From the centre of the cylinders to the centre of the crank axle is 10 feet 3½ inches. The bogie wheels are 3 foot 6 inch diameter, and the coupled wheels 6 feet 9 inches. The wheel-base of the bogie is 6 feet, from the bogie centre to the driving centre is 10 feet 2 inches, and the coupled centres are 8 feet 9 inches apart. The diameter of the boiler outside the largest ring is 4 feet 5 inches, and the length 10 feet 7¼ inches, the height of the centre line above the rail level being 7 feet 7 inches. The boiler contains 267 tubes of 1⅝ inch diameter. The length of the firebox casing is 6 feet 2¼ inches, the depth from the centre line at the front is 5 feet 3 inches, and at the back 4 feet 9 inches. The heating surface of the tubes is 1,227.34 sq. ft, that of the firebox 114.74 sq. ft, making a total of 1,342.08 sq. ft. The area of the fire grate is 18.73 sq. ft. The weight in working order is 42 tons 16 cwt, the bogie wheels carrying 14 tons, the driving 15 tons 8 cwt, and the trailing wheels 13 tons 8 cwt. The length of engine and tender over buffers is 53 feet 7 inches. The first three engines of this class, Nos 314, 315 and 316, have tenders of the same dimensions as the goods engines of the No. 433 class, but all subsequent engines have tenders slightly higher, carrying 2,600 gallons of water, and weighing, loaded, 26 tons 14 cwt. In the first three engines also the screw reversing gear stands much further back, extending beyond the cab pillar, but this has also been altered in subsequent engines.

Engines 316, 317, 318, 323, 324 and 207 are fitted for working foreign trains provided with the automatic vacuum brake. No. 206 was in the fatal accident at Wivelsfield on Christmas Eve, 1900, and is now fitted for burning liquid fuel. It is also noteworthy as having, on 2 October, 1898, taken the first 60-minute Pullman train from Victoria to Brighton. No. 208 is also fitted to burn liquid fuel.

In January, 1898, a four-coupled bogie express locomotive was built at Brighton Works similar to the standard type last described, but with a much larger boiler. This engine, which is numbered 213 and named *Bessemer*, *(Fig. 144)*. The frames, cylinders, wheels and motion are identical with those of its predecessors, but the boiler barrel is 9/16 inch thick, and made in three rings, the greatest external diameter being 4 feet 8 inches and the length 10 feet 7¼ inches. The height of the boiler centre is no less than 7 feet 11 inches. The barrel contains 299 brass tubes 1⅝ inches in diameter and 10 feet 8½ inches long between the tube plates. The length of the firebox casing is 6 feet 8¼ inches, the depth at the front below the centre line of the boiler 4 feet 11½ inches, and

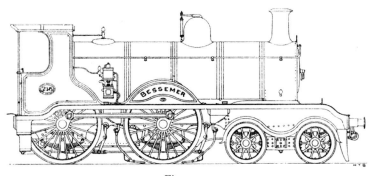

Fig. 144.

at the back 3 feet 6½ inches. The heating surface of the tubes is 1,354.47 sq. ft and of the firebox 110.35 sq. ft, giving a total of 1,464.82 sq. ft, and the grate area is 20.6 sq. ft. The total weight of this engine and tender in working order is 77 tons 14 cwt, the engine weighing 44 tons 14 cwt, distributed as follows: on the bogie wheels 14 tons 4 cwt, on the driving wheels 16 tons, and on the trailing wheels 14 tons 10 cwt.

To cope with the ever-growing traffic of the line and the increasing weight of the trains, a still further addition to the power of the locomotives soon became necessary. Consequently Billinton designed, in 1899, some extremely powerful four-coupled bogie express engines. The first of these, No. 52 *Siemens*, was built in December, 1899, followed by Nos 53 *Sirdar* and 54 *Empress* in January and February, 1900, respectively. These engines were designed to draw the heaviest passenger trains, such as that leaving Brighton at 8.45 a.m., which on Monday mornings is composed of the equivalent to 28 to 29 four-wheeled coaches, weighing, without engine and tender, about 348 tons. The cylinders of these locomotives are cast together in one piece, with the slide valves below. They are 19 inches in diameter by 26 inch stroke. To supply these large cylinders with steam a boiler 4 feet 10 inches in diameter, with a total heating surface of 1,635 sq. ft, and a grate area of 24 sq. ft is used. The blast pipe top is 5 inches in diameter, placed 1 foot above the bottom row of tubes. The boiler is fed by two combination injectors fitted on the back of the firebox. The reversing gear can be manipulated by hand or by compressed air from the Westinghouse air brake reservoir by a small two-way cock; no balance weights are required on the reversing shaft. The tender has a capacity of 3,000 gallons of water and six tons of coal, and is constructed with outside frames and bearings. The axle-boxes are fitted with doors in front, so that they can be easily examined. The tender is also fitted with hinged side plates that close on the sides of the cab, adding greatly to the comfort of the driver and fireman. The first three engines of the class, Nos 52–54, have the boilers made in three rings, the centre line being 8 feet 1½ inches from the rail; the steam dome is placed in the centre of the middle ring. All other engines of this class have the boilers in two belts, the centre line of the boiler being 8 feet 3 inches. The safety valves used on this class and the later batch of 5 foot radial tanks were designed by Billinton; the valves are direct loaded in two brass columns and cannot be tampered with by any person. The following tabular statement gives the principal proportions of this class:

Cylinders		Tubes No.	317
Diameter	1 foot 7 inches	Length	11 feet 4 11/16 inches
Stroke	2 feet 2 inches	Diameter	1⅝ inches
Centres	2 feet 1 inch	Grate Surface	24 sq. ft
Inclination	1 in 9½	H. S. forebox	126 sq. ft
Slide valves, lap	1 inch	Tubes	1,509 sq. ft
", lead	⅛ inch	Total	1,635 sq. ft
", travel	4 inches	Boiler pressure	180 lbs
", inclination	1 in 26	Wheels, bogie, diam.	3 feet 6 inches
Boiler.		", driving, "	6 feet 9 inches
Centre from rail	8 feet 3 inches	", trailing, "	6 feet 9 inches
Length	10 feet 10¼ inches	Centres of coupled wheels	
			8 feet 9 inches
Diameter outside	4 feet 10 inches	Weight, bogie	16 tons
Firebox Shell.		", driving	17 tons
Length outside	7 feet 7 inches	", trailing	16 tons
Breadth	4 feet 1 inch		
Depth, front	5 feet 2½ inches		
", back	3 feet 7¼ inches	Total	49 tons

The weight of the engine and tender in working order is 77 tons 7 cwt. The following is a complete list of the engines in this class, giving numbers, names, builders and dates:

No.	Name	Maker	Maker's No.	Date
42	*His Majesty*	LBSCR	–	May, 1902
43	*Duchess of Fife*	"	–	"
44	*Cecil Rhodes*	"	–	June, 1902
45**	*Bessborough*	"	–	"
46	*Prince of Wales*	"	–	July, 1902
47	*Canada*	Sharp, Stewart & Co.	4757	June, 1901
48	*Australia*	"	4758	July, 1901
49	*Queensland*	"	4759	"
50	*Tasmania*	"	4760	"
51	*Wolferton*	"	4761	"
52*	*Siemens*	L. B & S. C. Ry.	–	Dec., 1899
53	*Sirdar*	"	–	Jan., 1900
54	*Empress*	"	–	Feb., 1900
55	*Emperor*	Sharp, Stewart & Co.	4762	July, 1901
56	*Roberts*	"	4763	Aug., 1901
57	*Buller*	"	4764	"
58	*Kitchener*	"	4765	"
59	*Baden Powell*	"	4766	"
60	*Kimberley*	"	4767	"
61	*Ladysmith*	"	4768	"
62	*Mafeking*	"	4769	"
63	*Pretoria*	"	4770	"
64	*Windsor*	"	4771	"
65	*Sandringham*	"	4772	"
66	*Balmoral*	"	4773	"
67	*Osborne*	"	4774	Sept., 1901
68	*Marlborough*	"	4775	"
69	*Bagshot*	"	4776	"

* Fitted with liquid fuel apparatus on Holden's system.

** Fitted with Drummond's water-tubes in fire-box.

No.	Name	Maker	Maker's No.	Date
70	*Holyrood*	"	4777	"
71	*Goodwood*	"	4778	"
72	*Sussex*	"	4779	"
73	*Westminster*	"	4780	Oct., 1901
74	*Cornwall*	"	4781	"

No. 54, *Empress*, had the honour, on 2 February, 1901, of hauling from Fareham to Victoria, the special train conveying the body of the late Queen Victoria, *en route* for Windsor, while on 2 November, 1901, the same engine was used to convey the Prince and Princess of Wales from Portsmouth to Victoria on their return from their great Imperial tour.

On 21 December, 1901, No. 42, *His Majesty*, worked a theatrical special through from Victoria to Lime Street, Liverpool, over the LNWR, returning to London on Christmas Eve. On the same date, 21 December, No. 70 *Holyrood*, took the Pullman, Limited, from Victoria to Brighton, 51 miles, in 53½ minutes without a check.

The latest class of locomotive introduced by Billinton is a six-wheels coupled radial tank. These engines are designed for working fast heavy trains, which there is a great demand for during the summer months. This class is really similar to the standard 5 foot radial tanks before described, but is much larger and possesses a very handsome appearance. The six-coupled wheels are 5 feet 6 inches in diameter and the radial trailing wheels 4 feet. The cylinders, boiler (made in two rings) and wheel-base are the same as the 5-foot class, excepting the firebox, which is longer. The cab, bunker, and tanks are considerably larger and the boiler centre much higher. At the present time the following engines have been built:

567	*Freshwater*	Brighton Works	Oct 1902
568	*Carisbrooke*	"	"
569	*Kensington*	"	Nov 1902
570	*Armington*	"	"
571	*Hickstead*	"	Jan 1903
572	*Farncombe*	"	"
573	*Nutbourne*	"	Feb 1903
574	*Copthorne*	"	"
575	*Westergate*	"	March 1903
576	*Brenchley*	"	"

We have now completed our history of the locomotives of the LBSCR, having described and illustrated all those of which particulars are available, from the opening of the line to the present time, 1903.